The Online Courtroom

Leveraging Remote Technology in Litigation

Richard Gabriel and Ken Broda-Bahm
EDITORS

ABA
AMERICAN BAR ASSOCIATION
Tort Trial and Insurance Practice Section

Cover design by Jill Tedhams/ABA Design

Printed in the United States of America.

26 25 24 23 22 5 4 3 2 1

A catalog record for this book is available from the Library of Congress.

Contents

CHAPTER 3

CHAPTER 4

CHAPTER 5

CHAPTER 6

CHAPTER 8

Part Two

Best Practices in the Online Litigation Process

CHAPTER 9

CHAPTER 10

CHAPTER 16

CHAPTER 17

Introduction

At the time of the writing of this book, we have been in a period of great disruption. The coronavirus pandemic has claimed more than five million lives globally; millions have lost or changed jobs; there has been political unrest and growing fears over climate change; and millions of government institutions, organizations, and businesses have had to quickly change the way they operate to keep working during the public health crisis.

This crisis necessitated innovation. Chinese scientists sequenced the new COVID-19 virus in three weeks. Research teams in Boston, London, and Oxford developed a trial vaccine that was ready for testing in three months—a feat previously unheard of in vaccine development. James Dyson, an innovator in vacuum cleaners, designed a new ventilator in just ten days, while Ford, GM, and Tesla teamed up to produce ventilators for GE. A furniture company in Washington state repurposed its manufacturing capacity for mask production and shared its patterns and sewing processes with other factories. The UK's National Health Service built a 4,000-bed hospital in four days, and emergency rooms in Michigan doubled their ventilator capacity by adding a second tube so they could ventilate two patients at the same time.

The principle of stare decisis is based on a Latin phrase that translates as "stand by things decided"—the exact opposite of innovation. But the pandemic posed unprecedented challenges that modern jurisprudence had never seen. Namely, the justice system that was traditionally conducted in person, in a room where judges, lawyers, court personnel, litigants, and jurors (sometimes dozens) are all present and speaking, could no longer operate in the same way. Most courts struggled with this new dilemma: innovate or adhere to precedent.

I have spent the greater part of my 36-year-plus career working on jury trials. Early in the pandemic, before many grasped how long it would last, there was much discussion in the courts and the legal community about how these trials would have to be postponed until after the crisis was over, given that they involved many people in a room and the possibility of more exposure to the virus. However, given the mutating nature of virus variants and the uniqueness of the coronavirus, I understood it could be quite some time before the courts would be able to conduct business like normal. As a result, I started reading about innovation in the face of crises in different fields such as business or medicine. The main challenge I found in looking at applying innovation to the justice system was to develop methods for continuing access to the courts during the pandemic while not compromising the rights of litigants. So I was specifically looking for paradigms or principles that could allow continued court access without violating a constituent's legal rights. Not to emulate the exact procedures that we have all become accustomed to, not to replace in-person proceedings, and certainly not with the expectation that there would be a perfect solution to the problem. In studying innovation, I found that there were four basic principles that innovators from Thomas Edison to Elon Musk seemed to have employed to create new systems.

The first was to adopt a problem-solving mindset. In a crisis, there is a natural instinct in any organization to reduce risk—to contract for protection. We saw this with skyrocketing unemployment numbers throughout the United States in the early stages of the pandemic. But innovators tended to see disruption as an opportunity to actually expand and grow their organizations, allowing them to move from a static "wait and see" attitude to decisive action.

The second principle that struck me was the ability of innovators to see organizations, industries, or systems differently than they had traditionally been seen. Albert Einstein once was reported to have said, "We cannot solve our problems with the same thinking we used when we created them." This ability to "think different," to use an old Apple advertising campaign slogan, helps innovators to unstick hardened system patterns.

In a crisis, we have an instinct to protectively reduce our risk. To do so, we naturally look to established practices, procedures, and methods. Additionally, we are neurologically wired to behave based on historical data patterns and associations. Therefore, it is hard to see something differently from the way we have traditionally seen it. McKinsey & Co., one of the world's leading management consulting firms, polled executives of large corporations. An overwhelming majority stated they thought the pandemic would fundamentally change the way they did business in the next five years, but only a fraction of those executives felt they would be able to pursue this new direction in their businesses successfully. Making systemic changes is especially difficult for a legal system fundamentally based on precedent. So, whether in business or law, one of the challenges would be to overcome a negative mindset and to think about potential solutions without automatically dismissing them because of their unfamiliarity.

The third principle innovators have used in times of crisis is collaboration. While we sometimes have the romantic notion of a sole pioneer having a brilliant flash of inspiration or toiling away in singular isolation, most innovations are the result of collective efforts of teams with diverse points of view all working to solve a problem. While court systems tend to promote the individual discretion of judges in their own courts and defer to the local rules of a particular jurisdiction, this approach tends to create unique sets of rules in individual courtrooms and jurisdictions that can inhibit collaborative efforts. Yet, in this book, you will see how teams of judges and court administrators in western Washington state, northern California, Texas, New Jersey, and many other areas across the country worked to create systems that would provide temporary access to the courts for their constituents during the pandemic.

Finally, the fourth principle that is frequently used by innovators is learning and improvement through experimentation. Thomas Edison and his team of researchers supposedly tested more than 3,000 different lightbulb designs before coming up with their working model. This principle of experimentation is especially challenging for the courts, which are used to adhering strictly to the principle of precedent: proven methods or passed laws, statutes, or regulations before changing a procedure.

Keeping these principles in mind, I wrote an article for Law 360 in March 2020. It was a thought experiment about how the courts could conduct jury trials online. The article was written to postulate a way to keep the court system operational and accessible during the pandemic. As you will see in this book, as a result of the courts' adaptation to remote practices, advances in courtroom technology offer numerous promising future benefits to improve the justice system by providing greater access to the courts, greater representation for diverse communities, greater efficiency for numerous types of matters, and greater affordability for citizens who have difficulty acquiring legal services.

These benefits are important at a time where a majority of our citizens are losing faith in our justice system. Due to numerous factors—most notably an increasingly polarized political environment—both Gallup and the National Center for State Court polling showed a precipitous drop in the public's trust and confidence in the judicial branch of the government and the courts between 2020 and 2021.

In this book, we hope to illustrate how innovators across the country tackled and solved the problem of remote litigation; the advantages and the limitations of online courtroom procedures; the tools, technology, and skills needed for online courtroom presentations; and the current legal rulings on and legislative standing of remote procedures.

The perspectives in this book are not meant to advocate for replacing in-person with online procedures, but rather to augment current practices. We should not be afraid to challenge conventions, orthodoxies, and assumptions. We should not be afraid of change, even if it is initially uncomfortable. And we should not be afraid to ask "How can we improve?" if it ultimately provides greater access, economy, efficiency, representation, and a fairer form of justice.

Richard Gabriel,
Founder of the Online Courtroom Project

Acknowledgments

To make this endeavor possible, we would like to thank Sarah Forbes Orwig and Mike Beaver from the ABA publication team for their guidance and patience. We would also like to acknowledge the significant contributions of time and energy of all those involved in the Online Courtroom Project. We would like to thank the National Center of State Courts—and Paula Hannaford-Agor in particular—for their diligence in researching, tracking, and providing guidance for courts around the country during the pandemic. We also want to acknowledge and thank the Civil Jury Project at New York University for their ongoing commitment to jury trials and their thoughtful leadership in conceiving the advantages of online courtroom technology. We would like to recognize the Institute for the Advancement of the American Legal System for their ongoing commitment to improving the justice system. We also want to thank the National Institute for Trial Advocacy, which sponsored forums for the legal community to discuss the efficacy of online courtroom procedures during the pandemic, as well as the National Judicial College, ABOTA, the American Inns of Courts, and numerous other legal organizations for their sponsorship of those programs. We would like to recognize numerous innovative court systems—including those in western Washington state, Arizona, New Jersey, Texas, and Florida, as well as other individual jurisdictions around the country—for their leadership in providing access to the courts in such a challenging time. In particular, we would like to thank Judge Matthew Williams, Judge Marsha Pechman, Judge Pamela Gates, Judge Sean O'Donnell, Judge Glenn Grant, Judge Emily Miskel, and Judge Mark Drummond (Ret.) for their leadership in exploring online courtroom procedures. Finally, we would like to acknowledge all of the court personnel, judges, lawyers, legal support staff, trial consultants, technologists, and researchers for their tireless commitment in continuing to keep the courts open in such a difficult period.

Richard Gabriel and Ken Broda-Bahm

About the Editors

Richard Gabriel is president of Decision Analysis and founder of the Online Courtroom Project. He is a former president of the American Society of Trial Consultants and has written a book on his work in numerous high-profile trials and cowritten a book on jury selection. For 36 years, he has developed innovative solutions for clients and the justice system and is a frequent commentator on high-profile trials for CNN.

Ken Broda-Bahm, Ph.D., is a senior litigation consultant with Persuasion Strategies (a service of Holland & Hart LLP), in Denver, Colorado. He has provided research and strategic advice to clients since 1993. A member of the Advisory Board for the Online Courtroom Project and a past president of the American Society of Trial Consultants, he also writes extensively on topics of litigation persuasion and advocacy at the blog PersuasiveLitigator.com and frequently presents on effective courtroom communication.

About the Authors

Alicia Aquino is the CEO and trial presentation specialist with Aquino|LGW, based in San Diego, California. Ms. Aquino is on the ABA's LTRC Committee and the Women in Legal Tech Summit Committee and is well versed in trial preparation, mock trials, trial strategy, graphics, and the presentation of evidence. From the beginning of the pandemic, she has been actively assisting law firms with how to appear and present a case in remote and hybrid settings.

Theresa Wardon Benz is a partner at Wheeler Trigg O'Donnell LLP, in Denver, Colorado, where she represents clients in high-stakes appeals, litigation, and trials. She is an accomplished appellate practitioner with experience in multiple state and federal courts. Prior to joining Wheeler Trigg O'Donnell, she clerked for then–Judge Gorsuch on the Tenth Circuit Court of Appeals. In 2017, the American Bar Association named Ms. Benz one of the top 40 young lawyers in America in its annual "On the Rise" list.

Ken Broda-Bahm, Ph.D., is a senior litigation consultant with Persuasion Strategies (a service of Holland & Hart LLP), in Denver, Colorado. He has provided research and strategic advice to clients since 1993. A member of the Advisory Board for the Online Courtroom Project and a past president of the American Society of Trial Consultants, he also writes extensively on topics of litigation persuasion and advocacy at the blog PersuasiveLitigator.com and frequently presents on effective courtroom communication.

Ted Brooks has been providing trial presentation consulting services to law firms for over 20 years. He is one of the first and few to offer HIPAA-compliant Zoom hosting ("Technical Bailiff") for remote, in-person, and hybrid trials, arbitrations, and depositions. Specializing in complex matters, Mr. Brooks has trial experience that includes many high-profile and celebrity cases, including the Los Angeles Dodgers divorce and *People v. Robert Blake Criminal Defense*.

Lisa L. DeCaro is the cofounder of Courtroom Performance, Inc., a trial consulting firm dedicated to improving oral advocacy. Her practice specializes in witness preparation and trial strategy for the defense of civil cases. Ms. DeCaro is a national keynote speaker and the coauthor of the book *The Lawyer's Winning Edge: Exceptional Courtroom Performance.*

Chris Dominic, president and senior consultant at Tsongas, has practiced jury consulting for 23 years. He specializes in persuasion and social psychology, focused on the critical analysis of speeches and their impact on audiences. He leads the company's practice on cases ranging from everyday litigation to some of the highest-stakes trials in the nation. He is an active member and past president of the American Society of Trial Consultants.

Laura Dominic is a senior consultant with Tsongas and brings 24 years of experience in trial consulting to clients. Her expertise in trial strategy, strategic jury selection, and jury research spans most types of litigation on both sides of criminal and civil litigation. Since the outset of the pandemic, Ms. Dominic and Tsongas have offered innovative solutions for online jury research, jury selection, and jury trials.

Jeffrey T. Frederick, Ph.D., is president of Jeffrey Frederick Trial Consulting Services, LLC in Charlottesville, Virginia. He has assisted attorneys since 1975. He is a cofounding member of the Online Courtroom Project and member of the American Society of Trial Consultants and American Psychology/Law Society. He has written extensively on the topic of jury trials and trial advocacy, including three key books in the area.

Richard Gabriel is president of Decision Analysis and founder of the Online Courtroom Project. He is a former president of the American Society of Trial Consultants and has written a book on his work in numerous high-profile trials and cowritten a book on jury selection. For 36 years, he has developed innovative solutions for clients and the justice system and is a frequent commentator on high-profile trials for CNN.

Justice J. Gary Hastings (Ret.) graduated magna cum laude from Southwestern University School of Law in June 1972 and spent the next 13 years as a civil litigator. He was appointed to the Los Angeles County Superior Court in 1985 and then elevated to the California Court of Appeal, Second District, Division Four, where he spent the next 13 years until his retirement in 2006.

Marc King, Ph.D., is lead trial psychologist at Vital Justice, a consultancy he founded in 2007. In 28 years of trial psychology experience, he has worked on several of the nation's largest mass action lawsuits. He collaborated with Trialcraft for two years and now specializes in entrenched disputes in which litigants are akin to family. Dr. King's publications include "Bilingual Courtroom" and "Jury Perception," in the *Journal of Trial Diplomacy.*

A veteran business and real estate litigator and trusted advisor to bank general counsel, **Jeffrey Kirschenbaum** brings more than 30 years of experience to dispute resolution. He regularly appears as lead trial counsel in state and federal court and in arbitrations, as well as civil appeals. Admitted to practice in California, Mr. Kirschenbaum received his B.A. from the University of California, San Diego, and his J.D. from UC Berkeley School of Law.

Robert J. Kopka is a managing partner at Kopka Pinkus Dolin, a civil defense firm with offices in five Midwestern states. Representing nationally known corporations and insurance companies for more than 30 years, Mr. Kopka has argued appeals before the U.S. Supreme Court and handled cases from coast to coast. He sits on several boards and committees with his peers and is a founding member of the National Retail and Restaurant Defense Association.

Dr. Karen Lisko is with Perkins Coie LLP and specializes in courtroom persuasion. Dr. Lisko holds a doctorate in legal communication, a specialized degree attained by only a few, and is the author of two books, *Patently Persuasive* and *Proven Jury Arguments and Evidence.* She is past president of the American Society of Trial Consultants and is a founding member of the Online Courtroom Project, often advising the judiciary on remote jury trials.

Sarah Murray, MA, founder of Trialcraft, is an anthropologist and Fulbright Scholar with 20-plus years of trial consulting experience. She previously worked as technical writer and trainer, using her understanding of social systems to help people more happily and effectively integrate technology into their work practices. Ms. Murray has worked on thousands of civil and criminal cases and spent many hours in both mock and real courtrooms helping craft persuasive trial narratives.

Lisa Oberg, a partner with Husch Blackwell LLP in Oakland, California, centers her practice on providing strategic advice in high-stakes, volume litigation as national counsel defending thousands of cases throughout the country. She has worked closely with the courts on the creation and implementation of orders governing toxic tort cases, including those governing remote trials, and she provides insight and analysis on emerging trends in mass tort litigation.

Michael Pressman is an associate in the San Diego office of Sheppard Mullin. Previously, he was a research fellow at the Civil Jury Project at NYU School of Law and a law clerk to Judge Garaufis (Eastern District of New York) and to Judge Bacharach (Tenth Circuit). He holds a B.A. and an M.A. from Stanford University, a J.D. from Stanford Law School, and a Ph.D. in philosophy from the University of Southern California.

Michael E. Sandgren is a trial partner in the northern California office of Husch Blackwell, LLP, where his practice is focused on toxic tort, personal injury, and contract litigation. He has represented clients in trials from Monterey to Shasta County and has successfully defended corporate clients in some of the nation's most progressive jurisdictions. As a former extern with the California Supreme Court, he has written and spoken on litigation trends in California including publication in the *Los Angeles Daily Journal*. A pragmatic democrat, he is actively engaged in local politics and a member of the HB Political Action Committee.

Michael Elias Shammas is an instructor of law and Forrester Fellow at Tulane Law. After graduating from Harvard Law in 2016, he worked as a litigator at Paul, Weiss, Rifkind, Wharton & Garrison, LLP, for two years before completing two federal clerkships. Much of his contributions in this piece are informed by research completed as a research fellow at the NYU School of Law's Civil Jury Project.

Josh Splansky has been supporting high-stakes litigation since 2004. Since opening his own consulting and media practice in 2008, he is most often hired as a seasoned trial tech and to oversee visuals for litigation. After graduating as a Commonwealth Scholar from the University of Massachusetts Amherst, Mr. Splansky worked in TV and film production. He gained experience as a producer, director, writer, educator, and new media designer.

Geoffrey A. Vance is a partner in Perkins Coie's Commercial Litigation practice and firmwide chair of the E-Discovery Services & Strategy group, leading a team of more than 120 attorneys and technology professionals whose goal is to improve litigation outcomes and reduce the cost and risks associated with the discovery process. Mr. Vance is also the managing partner of the firm's Shanghai office and is on the firm's Asia leadership team.

Noah Wick, M.A., is the national director of litigation consulting at Trial Exhibits, Inc., where he advises trial teams on theme development, visual strategy, and impression management through creating demonstratives, medical illustrations, animations, and trial support, running trial presentation software. He is a national speaker on hybrid and remote jury trials, having consulted in both state and federal courts during the pandemic, including the first binding Zoom jury trial.

Part One

Challenges and Opportunities of Online Litigation

Chapter 1

Why Take Litigation Online? Disruption and Innovation

Richard Gabriel

In his Socratic dialogue, Plato wrote, "Our need will be the real creator," which then became the English proverb "Necessity is the mother of invention." 2020 was a year of disruption, need, and creation. The coronavirus, the economy, our divisive politics, and the social justice movement created a national crisis challenging the way we interact socially, the way we conduct business, the way we shop, the way our government works, the way we teach children, the way we take care of patients, and the way we view others that are different from ourselves. It challenged us to think differently about some of the most basic activities and values in our lives.

Our courtrooms and courtroom procedures are steeped in a rich advocacy tradition, aged over hundreds of years. Yet, in a single year, the pandemic challenged us to fundamentally change the way in which we conduct our litigation practices.

A Year of Disruption

As the coronavirus became prevalent in mid-March of 2020, all public gatherings were suspended as public health officials tried to better understand the nature of the virus and its implications on our daily activities and social interactions. Courts, prisons, government agencies, and law firms had to deal with significant administrative changes in their internal operations as well as maintaining an operating justice system. Even in the midst of a crisis, our constitutional rights as citizens are not suspended. This created significant challenges, as the science and medical communities were still trying to understand how the virus spread, and the courts attempted to balance public safety against every citizen's constitutional access to the justice system. Once the scientists, doctors, and our public health agencies gained a better understanding of the spread of the virus, the federal courts and numerous states implemented procedures to resume in-person proceedings with limited trials, extensive screening and cleaning practices, as well as masking, Plexiglas, and social distancing.

At the same time, lawyers and judges quickly adapted to many aspects of online litigation: depositions, hearings, mediations, arbitrations, and bench trials. Although there was a technological learning curve for the courts and counsel in these new arenas, the feedback from participants seemed to be generally positive. Judge Lisa Rau (ret.), the founder of Resonate Mediation and Arbitration in Philadelphia, had this to say about her mediation practice in the midst of the pandemic:

> In addition to safety, online mediations have brought unexpected benefits. Seeing people in their homes provides valuable insight as to who they are and makes up in some ways for the negatives of not being physically together. People feel more secure in their home than in a sterile conference room, so they are more confident and better at making decisions. Those involved in the decision making who ordinarily would not attend a mediation because of travel or scheduling limitations are able to attend part or all of the mediation because it is online. Online platforms provide unlimited real estate enabling people to meet privately in any number or configuration of breakout rooms. Sharing documents and doing calculations using a shared screen is easier. Long mediations are less exhausting since breaks are at home rather than in an alien office environment.[1]

In June of 2020, after more than three months in the pandemic, the National Center for State Courts conducted a national poll of 3,744 attorneys.[2] In that poll, 93 percent had a positive or neutral impression of remote hearings and 85 percent would recommend remote hearings to their colleagues or clients. However, only 43 percent said they were open to conducting some portion of a jury trial remotely.

At the same time that litigators seemed to be enjoying a relatively smooth transition in most of their online litigation practices, there seemed

---

1. Interview with Judge Lisa Rau (Ret.), Founder, Resonate Mediation and Arbitration (Mar. 29, 2020) (email question response).

2. Memo from GBAO, to National Center for State Courts, re: Jury Trials in a (Post) Pandemic World—National Survey Analysis (June 22, 2020), https://www.srln.org/system/files/attachments/2020%20Jury%20Trials%20in%20a%20%28Post%29%20Pandemic%20World%20–%20National%20Survey%20Analysis.pdf.

to be significant reluctance in entertaining the use of remote technology[3] to conduct jury trials. The worries about the use of online procedures in conducting jury trials seemed to be centered around a number of basic concerns:

- That attorneys would not be able to adequately observe jurors in voir dire to effectively select a jury
- That attorneys would not be able to observe jurors during the trial in order to gauge their reaction to the evidence and testimony
- That jurors would not be able to effectively judge the credibility of witnesses
- That jurors in lower socioeconomic groups would be underrepresented because they would not have the equipment or the internet capabilities to connect to the court
- That attorneys would lose the in-person connection that they would have with a jury and, thus, may lose their persuasive edge
- That jurors would lose empathy for the parties or the witnesses if they are not present in the courtroom
- That jurors would be distracted and not pay attention to the evidence and testimony
- That jurors would inappropriately research case issues on their computers or other devices
- That jurors would not take a court case as seriously as when they were in court and that the trial would lose the solemnity and weight of an in-court proceeding
- That remote trial proceedings would not satisfy the Confrontation Clause in the Sixth Amendment and make it difficult for attorneys to meet with incarcerated clients

Out of these concerns, there were calls on the civil side for more mediations, more bench trials, and a reduction in peremptory challenges and jury panel size. On the criminal side, there seemed to be greater flexibility on the part of prosecutors and defense attorneys to resolve cases through plea agreements to deal with the uncertainty of when trials would resume.

However, in the ensuing months, as it became more and more apparent that the pandemic would stretch on into the following year and the backlog of cases continued to grow, numerous organizations and courts started embracing innovation and testing various online trial formats.

---

3. During the course of this book, the authors will use both the terms "remote" and "online" to connote the use of technology in the courtroom

One of the earliest groups was the Online Courtroom Project (OCP),[4] a collection of retired judges, lawyers, trial consultants, trial technologists, and researchers who were formed to research and explore the possibilities and limitations of online technology in court proceedings. They conducted a full online demonstration trial from the jury summons process through verdict in June of 2020[5] and sponsored a two-day summit[6] with the National Institute for Trial Advocacy (NITA) called "COVID, the Court, and the Future of the Jury Trial" in November of that year. The OCP continued to collect data from around the country, to study the legal and social psychology literature on the efficacy of remote versus in-person presentations, and was involved in the state of Arizona's pilot project testing the difference between these two forums. Moving into 2021, the OCP wrote a white paper[7] on best remote courtroom practices and conducted numerous continuing legal education (CLE) offerings around the country, including a national webinar with NITA[8] where judges in Washington state, Florida, and New Jersey described what their jurisdictions and states had done to provide access to the civil justice system during the pandemic and their observations on the results of these remote proceedings.

The Civil Jury Project at New York University (NYU) School of Law also conducted a fully online mock trial in May of 2020[9] and published articles during the year on the legal parameters of remote trial procedures as well as conducting interviews with judges who were engaged in such

4. For more about the OCP, visit https://www.onlinecourtroom.org/ (last visited Nov. 6, 2021).

5. OCP, Online Jury Trials: Summary and Recommendations (2020), https://www.onlinecourtroom.org/demonstration-report.

6. National Institute for Trial Advocacy, *Day 1: COVID, the Court, and the Future of the Jury Trial*, https://www.nita.org/s/product/day-1-covid-the-court-and-the-future-of-the-jury-trial/01t4W00000DAR5vQAH (last visited Nov. 6, 2021).

7. Richard Gabriel, OCP & NITA, The Online Courtroom and the Future of Jury Trials (2021), https://www.onlinecourtroom.org/publications-and-manuals (last visited Nov. 6, 2021).

8. NITA, *The Online Courtroom Now and Post-Pandemic: Skills and Tools for Remote Advocacy*, https://www.nita.org/s/product/the-online-courtroom-now-and-post pandemic-skills-and-tools-for-remote-advocacy/01t4W00000CP0xhQAD (last visited Nov. 6, 2021).

9. For a short video overview of the project, see Civil Jury Project, *Highlights from the Civil Jury Project at NYU School of Law and CCCPC Virtual Mock Trial*, YouTube (July 21, 2020), https://www.youtube.com/watch?v=7U_aVbklGpU.

practices. They also wrote a memorandum that concluded that remote civil jury trials were both legally permissible and constitutional.[10]

During 2020, judges in Texas, Florida, Arizona, and northern California formed committees and task forces and experimented with remote trial formats in both civil and criminal trials. Perhaps the most forward-thinking area of the country was western Washington where both the state and federal courts studied and implemented detailed procedures to conduct both fully remote and "hybrid" trials, where jury selection was conducted online but trials were held in a convention center space or were completely online from jury selection through verdict. As mentioned, Arizona, always a careful innovator of trial reforms, conducted a series of trial tests, with assistance from the OCP, of different forms of in-person and remote trials and implemented fully online trials in parts of the state. Some of the federal and state courts conducting these trials required consent of the parties to participate in online trials and some did not. New Jersey was one of those states that first implemented remote trial procedures at the beginning of 2021, first by consent and then by mandatory assignment.[11]

Throughout 2020, many courts also implemented detailed in-person procedures for conducting trials. While many jurisdictions were able to implement some in-court jury trial practices, social distancing, cleaning requirements, the health concerns of jurors, as well as the shortage of courtrooms physically large enough to accommodate social distancing requirements severely limited the number of conducted trials to a fraction of what would normally be heard in a given year. This resulted in a massive backlog of both criminal and civil cases, which continued to grow as the pandemic continued. Some in-person proceedings also had to be delayed or mistrials declared after jurors, attorneys, judges, and/or courtroom staff became infected in the middle of the trial. Additionally, some cases also encountered legal appeals based on masking requirements for witnesses or attorneys in these in-person trials.

10. Michael Shammas & Michael Pressman, *Memorandum: The Permissibility & Constitutionality of Jury Trial by Videoconference*, Jury Matters (Civil Jury Project, New York, N.Y.), May 2020. https://myemail.constantcontact.com/May-Newsletter-of-the-Civil-Jury-Project.html?soid=1127815376566&aid=CVq1WlqBpgU.

11. New Jersey Courts, Virtual Civil Jury Trials during COVID-19 (2021), https://www.njcourts.gov/public/assets/virtualciviljurytrialscovid19.pdf?c=TA1.

In August of 2020, the National Center for State Courts conducted a poll of jury-qualified citizens in Texas.[12] In that poll, only 23 percent of prospective jurors said they felt comfortable reporting for jury duty in person and 43 percent said they would prefer to report for jury duty remotely, with the rest saying it did not make a difference to them whether they reported in person or remotely. However, more than 50 percent of citizens in larger metropolitan areas, Black jurors, and jurors with higher health risks stated they would prefer to report for jury duty online as opposed to in person.

At other points during the year, numerous webinars and CLEs were held by various bar and judicial organizations to discuss the experiences of participants in remote court and trial proceedings. Different bar associations and legal organizations published position papers on remote trial advocacy, mainly in opposition to the practices. In November of 2020, the two-day summit that was designed and implemented by the OCP and NITA included judges, court administrators, attorneys, trial consultants, trial technologists, and law professors, all discussing their experiences, opinions, and recommendations for best practices in this new environment. The summit was conducted in partnership with the National Judicial College, the American Board of Trial Advocates, the Civil Jury Project at NYU, the International Academy of Trial Lawyers, the American Inns of Courts, the Institute for the Advancement of the American Legal System, the American Society of Trial Consultants, the Professional Development Consortium, the Lawyers Club of San Diego, and Women Owned Law.

At the time of this summit, the OCP conducted a national poll of judges and attorneys about their experiences with remote proceedings. Similar to the poll conducted by the National Center for State Courts, most had a positive experience of their online court hearings, mediations, and arbitrations. Specifically, that poll tried to gauge the ability of the attorney and judge to communicate and manage the technology in an online environment.

In Figure 1.1, approximately two-thirds of the poll respondents stated they could communicate as clearly, if not more clearly, online than in an in-person meeting, while one-third felt they could not communicate as clearly.

12. GBAO, National Center for State Courts Texas Juror Poll (2020), http://www.litigationsection.com/uploads/NCSC%20Texas%20Juror%20Poll%20Results%20062320.pdf.

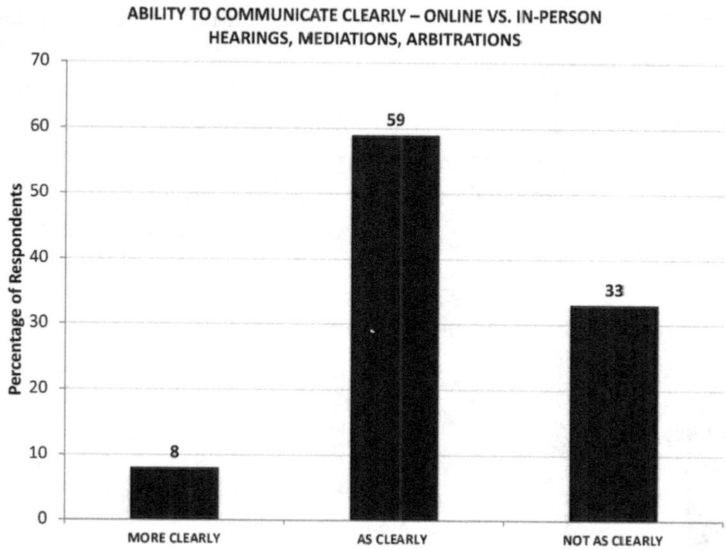

Figure 1.1

More than 80 percent of judges and lawyers felt it was very or somewhat easy to manage the technology in online hearings, mediations, and arbitrations, with less than 20 percent finding it difficult to use the technology, as illustrated in Figure 1.2. As these findings occurred in November of 2020, it would be interesting to see how these opinions may have

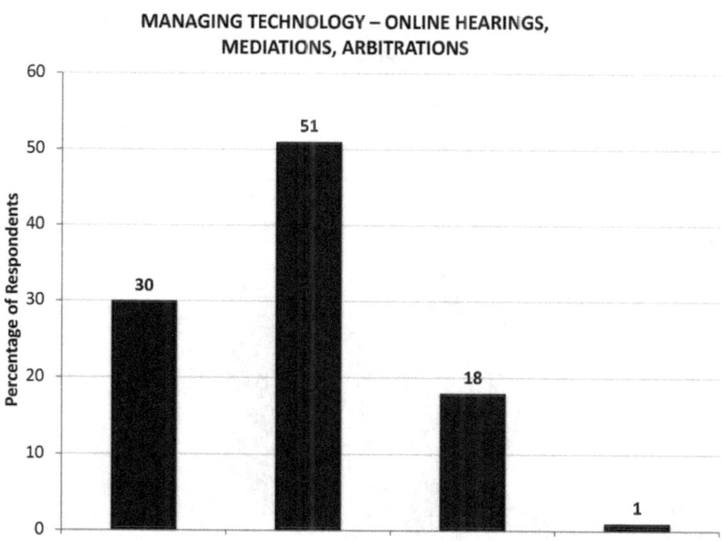

Figure 1.2

changed over time with greater frequency of use and familiarity with the technology in court proceedings.

Although far fewer had actually participated in an online bench or jury trial by November of 2020, those that had been involved in these new proceedings had a mainly positive experience (69 percent), describing that they were able to effectively communicate (68 percent) and manage the technology (75 percent) in those proceedings, as reflected in Figures 1.3 through 1.5.

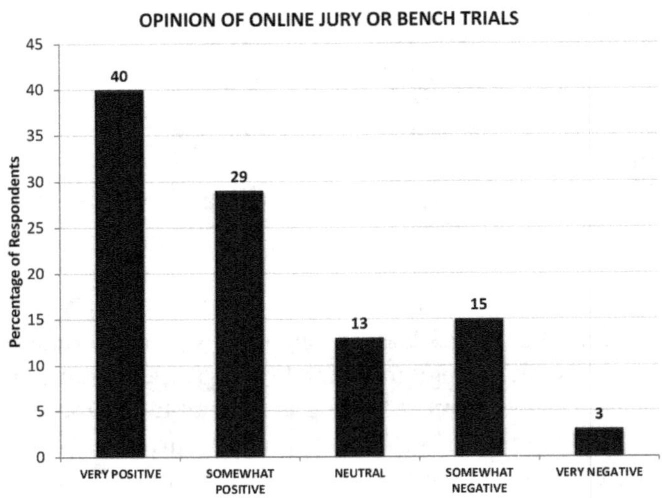

Figure 1.3

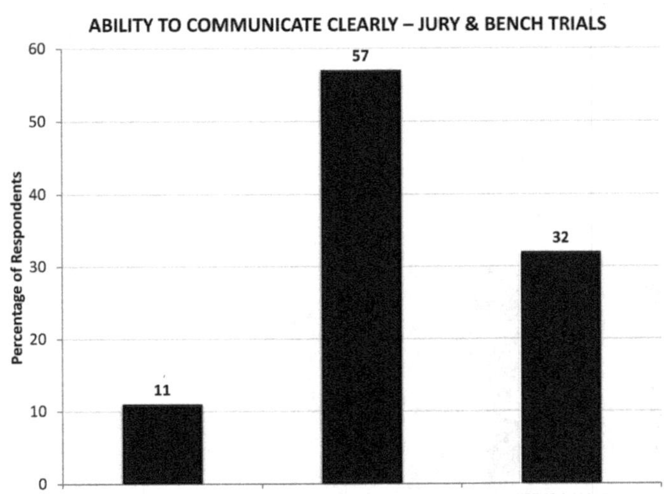

Figure 1.4

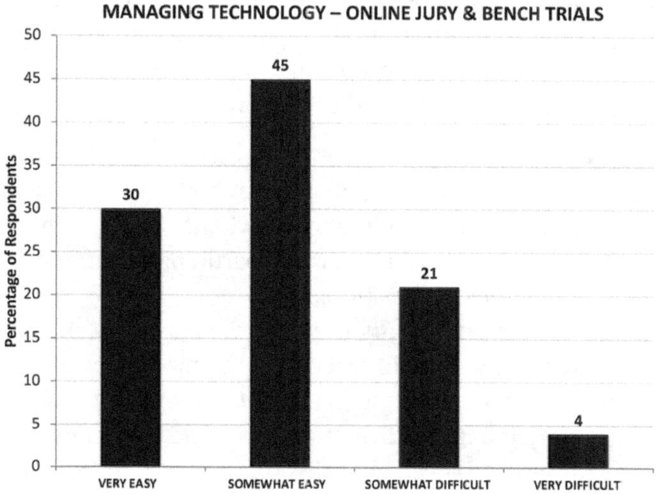

Figure 1.5

When those polled were asked about what they thought might be the benefit, if any, to conducting jury trials remotely (Figure 1.6), attorneys, judges, court administrators, and other justice system constituents who participated in the poll described a number of potential advantages.

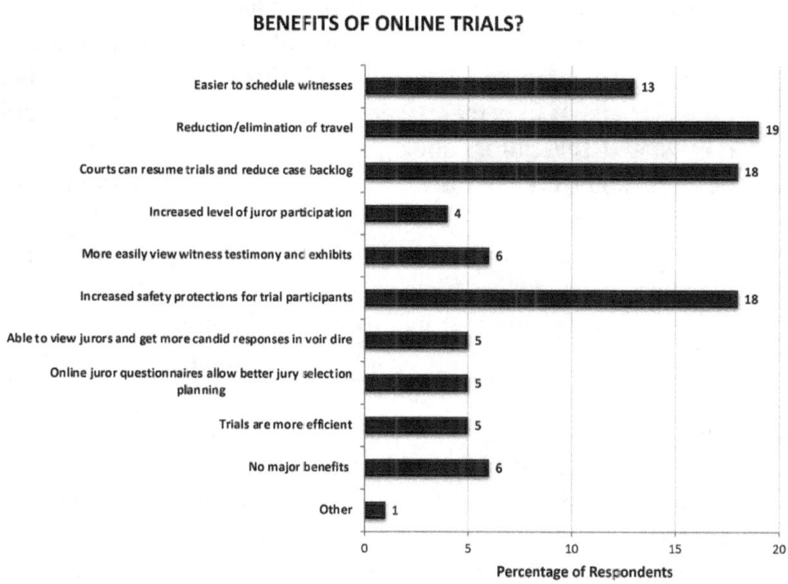

Figure 1.6

At the two-day summit sponsored by the OCP and NITA, there were 34 separate panelists and moderators in eight separate sessions. These panelists and speakers in dozens of subsequent webinars addressed all of the concerns previously listed about remote litigation, noting:

- Many courts commented how juror response rates had actually increased for remote trials compared to normal pre-COVID in-person jury summons response rates. Courts that conducted online jury selection allowed jurors to participate with smartphones, tablets, laptops, or desktops.
- Some judges remarked that their jury pools were even more diverse and representative of their communities because of jurors' ability and willingness to participate in remote trials.
- Most of the participants said they actually had a clearer view of individuals' nonverbal behavior, given that the jurors', witnesses', and attorneys' faces were right in front of them on the screen. As a result, jurors remarked about how they were able to easily judge witness credibility.
- Attorneys remarked that having a personal view of jurors inside their homes gave them useful insight into individual jurors, and being able to view multiple people on screen was often better than the profile view of jurors in the jury box in a courtroom. This was contrasted by participants in in-person trials who discussed the difficulty of being able to see the expressions of jurors or witnesses in a masked and socially distanced environment.
- Judges also remarked how attentive jurors were during their online trials. While there were inevitable technical problems, they stated that jurors took their job seriously, were easy to observe, and were not distracted by other devices or applications during the trial.
- While many attorneys reported a steep learning curve with the technology needed to present evidence and testimony in an online forum, jurors and judges reported being able to see evidence more clearly on a screen in front of them rather than having to look across a courtroom.
- Attorneys no doubt may have had more difficulty in establishing an emotional connection with jurors as they struggled to adapt to the new technology. However, *jurors* reported no problem in understanding the emotional content of cases or in experiencing a connection to the attorneys, parties, witnesses, or evidence.

- Judges have also observed that online legal proceedings, whether they be in the form of depositions, mediations, arbitrations, or bench or jury trials, are less rancorous and more civil, allowing fact finders to focus on the substantive issues and the parties to more easily resolve disputes.
- While the courts clearly did not want to undertake remote criminal trials, a review of case law did not conclusively show that there would be a Confrontation Clause violation in an online setting. In fact, appellate arguments were made that a masked, socially distanced environment may violate a defendant's Sixth Amendment right. The most significant obstacle cited to conducting criminal cases online was the ability of counsel to adequately prepare with their clients for trial.
- Judges have said that the ability to see criminal defendants in their homes with their families allowed them to have a fuller appreciation of the environment and life circumstances of those individuals, with a positive effect on their sentencing decisions.

Because of the uncertainty about the trajectory and length of the pandemic, the halting nature and health risks of conducting in-person trials, the tremendous backlog of cases, and the positive feedback of those that had participated in online trials, a number of states and jurisdictions developed task forces and implemented different forms of remote procedures. As of this writing, the following states and some federal courts are allowing or employing some form of online jury trial procedure:

Arizona	Nevada
Arkansas	New Jersey
California	Oregon
Florida	Pennsylvania
Illinois	Rhode Island
Kansas	Texas
Massachusetts	Washington
Michigan	Wisconsin
Minnesota	Some federal courts

Because of the relatively positive response to online trial procedures, Washington state has developed a Remote Jury Trials Workgroup and put together a best practices/frequently asked questions document to assist court constituents.[13] The state proposed legislation to make jury selection by videoconference an option for judges, even after the pandemic.[14] California passed legislation in September of 2021 to allow remote conferences, hearings, proceedings, and trials in civil cases.[15] Chief Judge John Tunheim in the District of Minnesota stated that he anticipated that they would keep holding remote civil trials even after the pandemic:

> It does take extra work . . . . But at the same time, it's a very, very convenient way to do a trial. For example, if I had a civil trial where I had lawyers from, say, Atlanta or San Diego or Seattle participating as primary lawyers in the case, and witnesses from all over the country, I might say, "Hey, we can do this a lot more inexpensively if you're willing to do it virtually. We can do this trial three weeks from now and get it done, as opposed to four or five months from now."[16]

The Family Law Judiciary, in a national survey, found that almost 64 percent of family law judges would continue hearing simple and uncontested divorces remotely and more than 60 percent of the judges felt that remote hearings promoted greater access to justice.[17] As one anonymous judge in the survey observed, "While we lose formality, we gain participation. Remote hearings let litigants, particularly self-represented litigants, participate more, and more participation equates to more access to justice."[18]

Additionally, the Arizona Supreme Court issued a report on post-pandemic recommendations[19] from their Plan B Workgroup, which met

---

13. Remote Jury Trials Workgroup, Washington Courts, Best Practices in Response to Frequently Asked Questions (FAQ) (2021), https://www.courts.wa.gov/newsinfo/content/Best%20Practices%20in%20Response%20to%20FAQ.PDF.

14. Washington Suggested New General Rule 41, https://www.courts.wa.gov/court_rules/?fa=court_rules.proposed (last visited Nov. 6, 2021).

15. S.B. 241, 2021–2022 Leg. Sess. (Cal. 2021), https://leginfo.legislature.ca.gov/faces/billNavClient.xhtml?bill_id=202120220SB241.

16. Cara Salvatore, *Minn. Judge Calls for More Zoom Trials—Pandemic or Not*, Law360, Mar. 30, 2021, https://www.law360.com/whitecollar/articles/1370514/minn-judge-calls-for-more-zoom-trials-pandemic-or-not.

17. Our Family Wizard, National Family Judiciary Remote Hearing Survey (2020), https://6a1ab614-8a16-459a-b02b-6cb58b4e4148.filesusr.com/ugd/850355_da459086fefc4e418845674916359c9c.pdf.

18. *Id.* at 11.

19. COVID-19 Continuity of Court Operations during a Public Health Emergency Workgroup, Arizona Supreme Court, Post-Pandemic

regularly during the pandemic. This workgroup surveyed judges, court commissioners, and courtroom staff. Forty percent of those polled said that technology-based platforms increased appearance rates. When asked about the benefits that litigants, attorneys, and other court constituents received by using technology-based platforms, survey respondents cited reduced travel time, taking less time off of work, reduced costs, increased safety, and increased ability to calendar hearings. One notable finding in this report was that failure-to-appear rates on eviction actions dropped from almost 40 percent in 2019 to as low as 13 percent by 2021. The report notes that "some Arizona courts are likely to continue to use technology to facilitate jury selection in some fashion,"[20] noting that there are significant efficiency gains by reducing the number of jurors that have to report to the courthouse, fewer claimed hardships, and jury selections that can be more quickly completed. In their pilot tests, Arizona found that participants in remote jury trials reported greater candor and an ease of attention during the trials.

Some of the Arizona survey respondents were quoted as saying:

- "We cannot and should not bring back hearings to in-person just because that's always how we've done things."
- "Litigants like [appearing remotely] because it reduces cost for travel time and time off work. Attorneys like it because it reduces the problems associated with being in multiple courts on any given morning."
- "To the extent possible, we should be seeing the court as a service and not a location."[21]

Most attorneys who have participated in online hearings, mediations, bench, and even jury trials have learned to effectively represent their clients and present their cases online. Yet even with the positive feedback from those that have participated in remote litigation, there has still been significant resistance and reluctance to engage in online procedures. Any new practice or skill is uncomfortable at first, but our personal comfort level should not determine what may be effective for our justice system. And tradition is comforting. It establishes guidelines, procedures, and routines for conducting our professional and personal lives, but it is important to distinguish tradition from habit. Habit is also comforting,

---

Recommendations (2021), https://www.azcourts.gov/Portals/216/Pandemic/2021/Post-PandemicRecommendations.pdf.

20. *Id.* at 24.
21. *Id.* app. at 37.

but it makes us resist new ideas and innovations that can actually improve the way we conduct court business and deliver justice. The challenge for anyone reading this book is the same challenge we give jurors: to set aside any preconceived beliefs that they may have about trials, to be fair and impartial, and to keep an open mind.

New Challenges, New Opportunities

The legal profession is no stranger to crisis. In fact, it is their stock in trade. Clients routinely turn to counsel and the courts to seek solutions to their conflicts. The pandemic has brought a new set of challenges to our practices, but also prompts a series of meaningful questions:

- Is our justice system reliant on a courthouse as a forum for parties seeking resolution for a civil or criminal dispute? In other words, is our system more reliant on a place than a process?
- Who does the justice system mainly serve, us or our clients?
- How adaptable and flexible are we to changing the way we have traditionally litigated cases?
- What changes in the litigation process are temporary fixes and what changes may become a more permanent part of our justice system?
- What resources, tools, and skills will the courts, counsel, and parties need to navigate changes that may become more permanent?
- How do some of the changes affect the role of advocates, fact finders, and witnesses in the litigation process?
- Will some of these changes actually provide an opportunity to improve the efficacy of our court system?

This book is designed to answer these questions—to provide research, observations, insights, and practical tips from judges, attorneys, trial consultants, and technologists who have participated in different forms of online litigation. While it is not the intention of the book to advocate for the replacement of in-person litigation practices and trials, it is the intention of the multiple authors to present options for courts and counsel to not only continue to provide the constitutional access to the courts in the midst of a health crisis, but to also look for ways to improve the efficacy of our courts and legal system in the future.

In the midst of the COVID-19 pandemic, extraordinary innovation flourished. While it usually takes years to develop a normal vaccine, researchers developed five viable COVID-19 vaccines in less than a year, due mainly to worldwide collaboration between scientists. They are not our only innovators; teachers created interactive lessons to keep their students engaged in remote classrooms, telemedicine skyrocketed, allowing routine medical appointments to be conducted online, and retailers, including grocery stores and restaurants, dramatically hastened their transition to online operations.

While the courts have long relied on precedent, during the pandemic, we all found ourselves in unprecedented territory and needed to seek unprecedented solutions. Our justice system was invented in 1791 when our founders, using the Magna Carta and the English common law system, literally *made up* our criminal and civil systems. Ever since then, new laws, new statutes, and new procedures have been debated, disputed, ruled on, and implemented in our courts. The justice system is designed to resolve disputes, both banal and unique, and over the years the courts have implemented new procedures and routinely decided cases of first impression. As a result, the legal profession and our justice system has frequently implemented and adopted reforms and innovations in order to adapt to changing laws and procedural demands. The following chapters in this book provide a number of potential advantages for considering online courtroom procedures.

Additional Future Benefits of Online Litigation

The advent of remote technology in litigation has a number of future potential advantages:

- The most prominent benefit is improved efficiency. The challenge of bringing multiple parties, witnesses, jurors, and counsel to a single physical location is an enormous logistical challenge and entails significant time and expense for courts, counsel, clients, witnesses, and jurors. Online litigation procedures obviously entail less travel time for all involved. Clients would no longer have to pay for counsel or witnesses to fly across the country to attend hearings, depositions, or trials. They could more easily attend depositions, hearings, or trials themselves without significant cost or disruption to their work schedules. Developing efficiency initiatives and measurements will provide long-term benefits for courts dealing with high case volumes.
- These efficiencies should also lower overall litigation expense. By lowering these costs, many litigants are provided greater

access to the justice system, which is especially advantageous for pro se parties.

- Jurors who do not have easy access to transportation and normally would not be able to attend trials could report for duty and could more easily arrange work schedules and childcare, depending on the court's trial scheduling. This would increase the diversity of jury pools and create a more representative cross-section of the community.
- Attorneys and jurors in rural areas of some states could more readily participate in litigation and trial without having to drive great distances.
- Jurors who have disabilities and may not normally be able to attend a trial because of the physical limitations of transportation or courthouse access would be better able to serve from their homes.
- Jurors who have a hearing or sight impairment may have better audio or visual clarity in a private location rather than in a courtroom environment.
- Those with English as a second language might be able to more readily participate if the parties and the court integrate and utilize remote translators that can translate for a single witness or juror without interrupting other jurors' listening. In the future, automated translation systems such as Google Translate may also be available to perform this function.
- The courts and the parties have more scheduling flexibility and witnesses may be more available to testify at prescribed times if they do not have the added difficulty of travel or commute time.
- Because the online forum encourages efficient communication, parties would be urged to streamline their cases to make better use of a judge's or jury's time. This would also allow the courts to manage their case load and resources more effectively.
- Because online litigation practice encourages calmer and less confrontational communication styles, it would allow a greater focus on case merit rather than posturing and personality. This could encourage a more amenable settlement environment for those cases where settlement is appropriate.
- Online courtrooms would also provide greater First Amendment access for the press and media who wish to observe a hearing or trial.
- Online litigation is also a great leveler. While courthouses carry both the authority and solemnity of our justice system, that same

environment can hinder participant comfort, candor, and communication. When we are all online, we are in our homes and offices, we all look pretty much the same, and we all have the occasional technical problems with the barking dog, the errant cat, or the wandering child. This humanizes us and makes us all more relatable. It enhances, not reduces, our investment in the people and the process.

A courthouse is made of wood, stone, steel, and concrete. So are classrooms, museums, theaters, and restaurants. Yet, even without the ability to physically gather in communal spaces, millions of students still learn lessons, we are still able to listen to music and see plays, dance, and art, and we can order takeout. Is it the same as our in-person experience? Of course not. Are there challenges to conducting litigation online? Of course. And are remote trials and procedures the solution for every case? Of course not. But the challenges and the novelty should not be a barrier to having a new set of tools and resources for the courts, clients, and counsel to more effectively manage the litigation process.

At its most basic level, the beauty of our justice system does not lie in its architecture, but in the elegant and dynamic process of human judgment: the process by which parties with disparate views reconcile their differences and resolve their conflict by impartial fact finders.

Chapter 2

Views from the Bench

Justice (Ret.) J. Gary Hastings

Summary

Simply stated, the American justice system exists to resolve disputes, whether arising from criminal action, family issues, commercial issues, personal injury, and so on, in a civil, timely, and orderly manner. Why should it matter what type of platform is used as long as the dispute can be resolved to the satisfaction of those involved? And that is where the focus should be.

Among others, constitutional issues such as due process, the rights of confrontation, and public trial play into the equation. As do issues relating to witness credibility, the ability to effectively introduce evidence, selecting and impaneling unbiased and representative juries, and ensuring that jurors remain focused and return verdicts that comport with the law and the facts. In addressing whether remote hearings and trials are appropriate, some of the issues are purely legal in nature while others are procedural. But procedural issues can have an impact on due process concerns. Many of these issues will ultimately be addressed by the appellate courts in the future. But given the benefits of remote hearings and trials as discussed in this chapter and overall in this book, there is no doubt in my mind that remote hearings and trials are here to stay.

What we need to address are the types of hearings and trials that lend themselves best to the remote platform and how to ensure that due process is provided. I attempt to provide a judicial perspective herein.

Procedural Justice

In the Winter 2006 volume of *California Courts Review*, Tom R. Tyler, PhD, published an article titled "What Do They Expect?," which reported on studies done to determine what people want from the courts. His conclusion was that people are more interested in how their cases are handled, which he called "procedural justice," than whether they win their case, which he called "distributive justice." He wrote:

> The idea that people might be more interested in how their cases are handled than whether or not they win often strikes people as counterintuitive and wrong-headed. Yet it is the consistent finding of numerous studies conducted over the last several decades, including a recent study of California state courts. These studies show that people use ethical criteria to evaluate their experiences, and that they particularly focus on their views about appropriate ways for authorities to act when deciding on how to resolve legal problems.[1]

He identified four factors that need to be addressed in aid of procedural justice: (1) voice—people want to be able to explain their case to legal authorities; (2) neutrality of authorities—they want the authorities to make decisions based on consistently applied legal principles and the facts of the case, not personal opinions and biases; (3) respectful treatment—they want to be treated with dignity and politeness; and (4) trust in authorities—people react favorably to the judgment if the authorities are benevolent and caring and are sincerely trying to do what is best.[2]

The American Judges Association has embraced the concept of procedural justice, which they term "procedural fairness." It is within this context that I provide the following discussion relating to my opinions about remote hearings and trials.

Hearings

Trial Courts

The vast majority of cases filed are usually disposed of without trial. But numerous hearings are conducted in these cases by judicial officers, which may have a significant impact resulting in how the case is ultimately resolved and may affect the perception of the public and the participants. Many courts now are handling these pretrial hearings in civil and criminal matters remotely even when witnesses are being called and evidence is being presented. The convenience and efficiency for the

1. Tom R. Tyler, *What Do They Expect?*, CAL. CTS. REV., Winter 2006, at 22.
2. *Id.* at 22–23.

court, parties, and attorneys is especially appreciated when the matter is a simple status conference or routine law and motion hearing that does not take more than five or ten minutes. But even when the hearings are more complex and require witnesses and exhibits, longer hearings can be successfully handled remotely.

Judge Kimberly C. Priest Johnson presides in the Eastern District of the U.S. District Court in Texas. She wrote an article for the *Bencher*, the magazine of the American Inns of Court, titled "Virtual Open Court."[3] She described her experiences in conducting remote civil and criminal hearings during the pandemic. One of her hearings was on a patent case involving attorneys from Delaware, California, Washington, D.C., Illinois, and Texas. The parties exchanged exhibits beforehand and all appeared remotely, resulting in what she termed an effective oral argument. She also heard a remote motion for preliminary injunction where counsel appeared from throughout the United States, live witnesses appeared from around the world, and exhibits were again exchanged and used during the hearing. Court staff had a trial run before the hearing and the actual hearing proceeded smoothly. Finally, she described a remote detention hearing involving four defendants all of whom were detained in local jails. There were four separate defense attorneys, a federal prosecutor, a federal agent, four separate witnesses, the court, and its support staff. Everyone appeared remotely. Despite the fact the assorted connections varied, she was able to conduct the entire hearing and hear from all parties over a few hours' time. She believed each defendant was afforded the same caliber of hearing as would have been provided in her courtroom.

The three cases described demonstrate how remote hearings can save time and money for the parties and attorneys and be more efficient for the court. There was no need to arrange for travel and lodging for all participants or for the attorneys to charge the clients for the travel time and accommodation expenses. The court and counsel were able to arrange a schedule for the timing of the witnesses and exchange of documents and exhibits to be used during the hearings.

In the early months of the pandemic a number of counties in the Bay Area of California imposed severe restrictions on businesses and the courts, Alameda being one. In March, the court closed operations due to the dangers of COVID transmission. I learned that Commissioner

3. Kimberly C. Priest Johnson, *Virtual Open Court*, BENCHER, July/Aug. 2020, https://home.innsofcourt.org/AIC/AIC_For_Members/AIC_Bencher/AIC_Bencher_Recent_Articles/2020_JulAug_Johnson.aspx.

Michael Bishay began handling complex family law settlement conferences remotely so I contacted him and did an interview over Zoom. He shared his thoughts with me.

When Commissioner Bishay approached the court administrators about handling his settlement conferences remotely, they were receptive but told him that under the circumstances they could not guaranty he would have a clerk or any effective backup. He decided he could handle the technological and other issues himself. The court provided access to its BlueJeans platform and the commissioner set out to arrange for and conduct settlement conferences, sending out notices and making all arrangements himself. He found attorneys were willing to work with him and advise their clients to participate in an attempt to settle their cases and avoid the delays that would result from waiting for courtrooms to reopen. While his success at settling cases prior to the COVID crisis had been excellent, he believes the percentage has risen. The platform allows him to assign the attorneys and their parties to breakout rooms that he controls and can visit whenever he feels it necessary. If others are involved, for example valuation experts, he can provide breakout rooms for them to confer. The parties, attorneys, and others are more relaxed being in familiar surroundings and do not have to suffer the commute to the court and find parking. There often were not enough conference rooms in the courthouse to accommodate everyone, meaning they would be shuttling in and out and having to sit on cold concrete benches while the process played out. And they seem more civil and receptive being in their own surroundings. He has found that face-to-face communications over the monitors is oftentimes better than in-person communications in the courthouse. He is more productive and able to work during times when the courthouse would normally be closed and believes that lawyers are finding they can be more productive and charge the clients less.

Commissioner Bishay also oversees a virtual self-help status conference program that is solely used by low-income pro per parties who need help from staff attorneys and volunteers in working their way through the system. It is a partnership between the court's Self-Help Center and a local legal aid organization and funded primarily by a grant from the state, and is held once a week with up to 20 cases set each day. The parties are provided notice and instructions on how to use and sign into the remote platform. When this program was held in the courthouse, each party had to wait his or her turn to be assisted because of the limited availability of conference rooms. Now, several parties are able to receive one-on-one assistance from staff or volunteer attorneys simultaneously using the unlimited virtual breakout rooms. This method saves time for litigants, staff, and volunteers. At first, they expected only a few parties would use the system but it has turned out to be very popular. The parties have the opportunity to appear visually by computer, tablet, smartphone, or audio only by phone. They discovered more people have access to computers and smartphones than they initially expected. The parties are happy because they can appear from home or work without having the commute that would otherwise be required. And the judges and administrators are pleased to be able to keep the cases progressing through the system.

Commissioner Bishay also reviews stipulated family law judgments. Some require a short prove-up hearing. In the past, attorneys and litigants had to block at least half a day and incur the cost associated with attorney fees, travel, parking, as well as loss of work just to conduct this short hearing. Now, these hearings are conducted in just a few minutes using BlueJeans with incredible cost and time savings.

Appellate Courts

Appellate courts differ from the trial courts in one important aspect: no evidence is taken and no witnesses appear. The appeal is decided based on the record established in the trial court and arguments presented by the parties in their briefs. Oral argument is the only participating part of the appeal and it is often waived in the more mundane cases. Thus, technology does not have to be as sophisticated as in the trial court. But that is not to say it is unimportant.

When the pandemic hit, most appellate courts in California and elsewhere turned to remote hearings. We interviewed Presiding Justice Arthur Gilbert of the Second District, Sixth Division, and Thomas Willhite of the Second District, Fourth Division, to find out how they perceive remote oral arguments. There is a consensus that while virtual arguments can be effective and save travel time for the lawyers, the justices miss the human element of an in-person hearing. But they recognize the efficiencies offered by virtual hearings and both agree they are most likely here to stay.

Virtual hearings cause some problems of which the justices and the lawyers must be aware. Both pointed out that lawyers can be a bit too relaxed, as can the justices. Lawyers should be warned that at all times they can be seen on the screen even when they are not arguing, and the justices must also keep that in mind. Distractions can arise when one or more of the lawyers or justices do something on-screen that makes it appear that complete attention is not being given to the case. Asking questions can also disrupt the flow of an attorney's argument and thoughts, although that occurs during in-person hearings as well. But when remote, there is often a bit of a delay, which can throw things off. The justices should consider setting rules about how and when questions are asked.

The discussion turned to the record on appeal when considering remote trials. What will the record be? Will it be transcribed from the virtual recording, will a court reporter still be involved, or will the justices have to review the video recording of the trial? Each was concerned that if they had to review the video record of the actual trial, issues of witness credibility, which are not usually addressed by the justices, will be called into play.

The bottom line is that each agreed that remote hearings are here to stay. But they preferred in-person argument if there is no hardship to the parties or attorneys from having to travel to the courthouse.

Bench Trials

Bench trials are similar to more complex pretrial hearings, which require evidence and witnesses. But a bench trial is just that, a trial that will determine the ultimate outcome of the case. For that reason, counsel, the parties, and witnesses may take a different approach than a pretrial hearing. But the judge's role is the same: to provide a complete and fair hearing or trial and decide the issues impartially and according to the law.

Commissioner Bentrish Satarzadeh handles traffic cases for the Alameda County Superior Court remotely. I contacted her and she agreed to be interviewed over Zoom. After the courts closed, she was able to return to the courthouse in May of 2020 where she handles the daily traffic calendars on the BlueJeans platform. She learned that court administrators were not able to guaranty that she would have the necessary support of clerks but she has learned to handle the platform herself so she does not need anyone to help her. She reports this has been a boon for those who appear pro per because they do not have to drive to the courthouse and wait around for their case to be heard. Instead, they can appear from home or work and be productive while waiting for their case to be called. Commissioner Satarzadeh uses the gallery mode and has all participants appear at the beginning of her calendar just as in the actual courthouse. She runs through the calendar and sets up an estimated order for the cases. She then tells the participants approximately when their case will be heard and allows them to continue doing what they are doing at home or at work but they must continue to be on the monitor and keep abreast of what is happening so they will be ready when she calls their case. At first she was worried the parties would not waive personal appearances but that turned out not to be a problem; almost all people appearing pro per give permission. And she has seen more not guilty pleas than when the parties were required to appear in person. She attributes this to the convenience of the parties to have their trials conducted virtually. She opines that handling matters virtually has provided up to 99 percent access to justice in these types of cases. She also believes that her communication with the participants is much better and even a bit more intimate than in the courthouse setting. This provides better insight into the facts of the case and the participant when she needs to choose a sentence that best fits the circumstances. Participants report back to her that they are satisfied with appearing virtually and they want the court to continue the program after the pandemic has subsided.

Judge Jeffrey Ross of the San Francisco Superior Court has tried a number of bench trials remotely. I spoke with him over the phone and he sent me an email with his thoughts, which I summarize as follows:

- Because witnesses must wear masks at all times while in the actual courtroom, witnesses can testify remotely using Zoom without masks and thus allow the finder of fact to see their expressions. He believes this approach has been especially effective in cases in which the witness is assisted by an interpreter. The speaker and interpreter can see each other—unmasked—on Zoom, which facilitates communication. It also improves communication with the court reporter since a masked witness in the courtroom may be harder to understand than an unmasked witness testifying via Zoom.
- One impediment of Zoom is the challenge of speakers interrupting one another. He resolved that issue during testimony by asking lawyers with objections to raise their hand so that the witness will see that there is an objection before answering the question.
- He believes another helpful technique is to have one person orchestrating the use of the Zoom technology. In a court trial where a dozen people were participating in the trial, the technician "pinned" the judge, witness, and lawyer questioning and objecting. As a result, it was easy for the witness to see the attorney raise his or her hand to object, which avoided interruption. The technician also managed the screen sharing to allow the witness, court, and counsel to view exhibits. The result was faster examination than in court, where lawyers approach the witness to provide documents to review. In addition to having the exhibit

on the screen, the witnesses testified from the lawyer's conference room where a complete set of exhibits was also available.

Judge Ross praised the independent technician who was hired and paid for by the parties for keeping matters running smoothly. We, the Online Courtroom Project (OCP), have found that a so-called "technical bailiff" is essential for virtual trials whether it be a court employee trained and assigned for that purpose or an independent technician brought in by one or more of the parties.

I have also been in touch with Judges Matthew W. Williams and Sean O'Donnell from King County Superior Court in the state of Washington. Their court has handled in excess of 600 remote bench trials over the past year, which has helped cut down on the backlog created by the pandemic. They agree that using a remote platform has expanded access to justice, especially where self-represented litigants are involved. Those litigants do not have to take time off from work or obtain childcare and travel to court. As Judge Williams noted when I spoke with him: "There are more people in America who have cell phones than cars or bus passes."[4] And they agree that it is cheaper for parties when lawyers do not have to bill for travel time or time wasted in court waiting for their cases to be called. And lawyers have more flexibility to handle matters for other clients during the time they would have been traveling to and being at the courthouse.

Jury Trials

The Debate

When I spoke with Judge Ross of San Francisco he told me that the judges in his court are reluctant to do jury trials remotely. He said that some judges have expressed concerns that allowing jurors to participate in trials remotely using Zoom or BlueJeans would interfere with the court's ability to assure that jurors are attending to the proceedings and not engaging in prohibited conduct during trial. For example, a juror may appear to be paying attention to the trial but may be looking at a phone or a computer screen without the judge knowing it. A juror may be using a computer during trial to research issues related to the case. While courts cannot prevent misconduct, allowing jurors to participate remotely increases the possibility for mischief.

4. Interview with Judge Matthew W. Williams, King County Superior Court, Washington.

On the other hand, judges in Alameda County have conducted completely remote jury trials with success, which I discuss a bit later in this chapter. Judge Williams from King County, Washington, reports that as of April 26, 2021, his court has tried more than 111 civil jury trials start to finish. The judges involved are very happy with the way these jury trials have been handled and resolved. They have devised means for monitoring the jurors and developed jury instructions to address the concerns the San Francisco judges have. Jury instructions are discussed in Chapter 18 of this book.

While virtual criminal trials offer the same challenges as civil trials, there are two very important constitutional concepts to consider: the rights of confrontation and to be effectively represented by counsel. If the defendant and his or her counsel are each in a remote location, how are they going to communicate effectively and securely? And does a remote trial where the defendant and witnesses are in separate locations satisfy the Sixth Amendment right of confrontation? Of course to the extent that a virtual trial may impede these rights, the defendant can agree to proceed after a full disclosure of the potential problems. But if defendants do not agree and courts move ahead with virtual trials, these issues, if preserved, will be addressed by the appellate courts. Until we receive guidance from the appellate courts, judges must be attuned to these issues and do the best they can to provide due process.[5]

The OCP Demonstration Civil Jury Trial

In June of 2020, the OCP put on a demonstration virtual civil jury trial to see how it would play out. It was a mock slip and fall case that had been patterned after a real trial. I presided as the trial judge from my home. Practicing lawyers from different cities remotely represented the parties, an injured plaintiff and a nationwide drugstore chain. Witnesses appeared remotely as did volunteer jurors who were recruited by members of the OCP from throughout the nation. The jurors were sent preliminary questionnaires asking information about their technical expertise for participating in a remote trial. Those who were not able to guaranty they would be able to adequately participate were released. For those who remained, follow-up questionnaires for use during voir dire were then sent dealing with issues of interest to the parties and counsel.

5. See *State v. Sweidan*, 13 Wash. App. 2d 53 (Wash. Ct. App. 2020), where the issue of having a witness testify remotely was discussed, and *Missouri v. Smith* (2022 WL 105131), where a criminal conviction was reversed by the Missouri Supreme Court because the trial court permitted a witness to testify remotely (https://law.justia.com/cases/missouri/court-of-appeals/2021/ed108626.html).

The first day of trial was used for jury selection with traditional voir dire, challenges for cause and use of peremptory challenges, and it was processed on an Adobe platform. After the jury was selected, counsel made opening statements. We then recessed until the next morning. The next day we used the Zoom platform and heard from four witnesses, with the use of exhibits, instructed the jury, and had closing arguments. The jury then deliberated for an hour as we watched. The jury ended up divided on the outcome. We recessed the trial and interviewed the jurors, attorneys, and witnesses about the experience.

My thoughts about the trial and the process are as follows:

- Overall, other than the platform and technical issues, the trial proceeded similar to a courtroom trial. I had no difficulties dealing with the attorneys, the witnesses, or the jurors. Each party was able to present its side of the case and the jurors, with one exception, were engaged and focused on the appropriate facts and law during their deliberation.
- I preferred the Zoom platform to the Adobe platform. However, both platforms were being adapted for use during the trial and need to be further developed for such use. From the judicial perspective I would want to have all jurors visible at the same time and all of the time on my screen along with the attorneys and the witnesses. It is not necessary that I be on the screen. In a real courtroom that is what I can see and it helps me monitor the jury to make sure they are paying attention. I had to scroll through the bottom of my screen to see them and I missed a juror who fell asleep. I do not think jurors need to be able to see themselves on their screen during the trial but they do need to see the judge, the lawyers, and the witnesses. And lawyers should be able to see the judge, the jurors, the witnesses, and themselves. Jurors do need to see each other during deliberations.
- Regarding participants, obviously a judge is needed. We had two "technical bailiffs." One made sure that witnesses were available, were ready to proceed, and brought them in from the breakout room when needed. She also provided a breakout room for sidebar conferences between me and the attorneys. The other one aided the attorneys in their use of the exhibits. A court clerk is needed to keep the minutes and the exhibits and probably should not be given any technical role in addition. Depending on the platform used, with the ability to record the proceedings, a court reporter may or may not be necessary as long as the record can be appropriately certified.

- Jurors must have the right equipment to be able to participate without technical glitches. That will require more work at the beginning for court staff to make sure all potential jurors will be able to adequately participate. Monitoring the jurors is a bit more problematic than in the actual court setting. But if they are properly admonished and observed there should be no problem. During my years on the trial bench I found that on many occasions jurors would appear to lose focus and even fall asleep—usually the result of less than compelling advocacy on behalf of the attorneys.
- I thought the witnesses were more intimate online as compared to the court setting. And I did not see anything that may suggest deciding credibility would be different than in the court setting. In fact, jurors who had previously been in courtroom trials commented after this experience that they felt they were able to judge the credibility of witnesses better than in the courtroom setting.[6]
- Documentary exhibits should not be a problem as long as the attorneys and other personnel are prepared to launch them on the platform and the platform can handle them. I can see a problem with nondocumentary exhibits to some extent. But I think that video and photographs can be used effectively to depict this type of evidence, much like use of the video during our trial.
- Use of exhibits with the witnesses, as long as planned ahead, should be no problem either, whether during direct examination, cross-examination, or for impeachment. Document sharing must be addressed before the trial at a pretrial conference so that it works seamlessly.
- I saw no difference in dealing with attorneys online as compared to in-court proceedings. Sidebar conferences were easily handled through the technical bailiff when needed. The attorneys will need to be prepared to use the platform and a trial run-through should probably be handled ahead of time.
- Some jurors were concerned that they could not vote anonymously. Zoom has a feature that does allow such voting but we did not use it for this trial. I have been involved with meetings where votes were taken by Zoom and it worked fine. When a vote is called for, the technical bailiff would put up the questions

6. Volume 51, Issue 1 of the Southwestern Law Review (2021) provides two articles addressing witness credibility and remote hearings: Susan A. Bandes and Neal Ferguson, *Empathy and Remote Legal Proceedings*; and Karen Lisko, *Bearing Witness to, Well, Witnesses: An Examination of Remote Testimony Versus In-Court Testimony.*

and jurors would then choose the options. The technical bailiff would then provide the tally to the jurors to see how many voted for each choice without disclosing who voted which way.

- Regarding the record on appeal, I have learned that a digital record is much easier and more convenient to use than the traditional clerk's and reporter's transcripts. However, if the record will include actual video of the witnesses, a problem could arise relating to how the appellate court addresses credibility issues. At the present, most appellate courts do not independently review witness credibility, leaving that issue to the trier of fact absent clear evidence that the witness was in fact lying.

Thinking back over the trial, I believe that each of the four elements identified by Tom Tyler in his article was accomplished. The parties were able to present their side of the case effectively, the jurors, or authorities who attempted to decide the case, appeared to be applying the law and the facts in a neutral but informed manner, and everyone acted respectfully. The facts were such that jurors were able to rationally come to different conclusions based on the law involved. I do not believe the outcome would have been any different if the case had been tried in an actual courtroom.

The Alameda County Experience

Alameda County Superior Court was the first court in California to conduct a remote civil jury trial. It was a three-and-a-half-month trial involving a claim of mesothelioma caused by asbestos exposure and was presided over by Judge Brad Seligman. He was interviewed and stated that before he conducted this trial he was a "complete skeptic" of virtual trials. But his experience with this trial changed his mind.

Judge Seligman's first concern was being able to obtain a cross-section of potential jurors who would have the technical ability to participate. It turns out his concern was unwarranted. The summons resulted in returns equal to or a little greater than for in-person trials. And the vast majority of prospective jurors had the technical capability to participate. He only lost a few for cause and can recall losing only one for technology issues; the prospective juror's home was to be without electricity for at least one day because of an electrical shutdown to make repairs. Jury selection was relatively easy. He utilized his normal six-pack approach and conducted challenges at a virtual "sidebar" conference. He and the lawyers were able to see the jurors clearly on the monitor and he felt that he was able to get a better view of them than in the actual courtroom. He said the same was true for the witnesses who testified. And the jurors and witnesses were not required to transport themselves to the courthouse and try to find a place to park. Afterwards the jurors told him they liked the experience of serving on a virtual trial.

A virtual trial requires more staff than in-court proceedings, at least one clerk and two others. One was needed to keep an eye on the jurors at all times to make sure they were paying attention and had not lost connectivity. He also believes that the trial lasted probably 25 percent longer than an in-court proceeding due to shorter hours and technological problems that needed to be addressed. He did say that the defense lawyers made quite a few objections to the fact they were proceeding virtually and there were at least three writs taken to the court of appeal, none of which were successful. One time an objection was made when one of the attorneys saw a cat on the lap of one of the jurors. The objection was overruled. He does not believe the outcome would have been any different if the trial had been conducted in the courtroom.

While Judge Seligman's trial was underway Judge Jo-Lynne Lee from Alameda County Superior Court started and presided over a second asbestos trial. We asked if she would share her experience, pros and cons, about the trial. She provided us with written comments that I summarize as follows:

Preliminarily she notes that much depends upon the resources of the individual court to handle remote trials. For instance, do they have sufficient court staff to monitor jurors at these trials? Do they have sufficient court staff with Zoom (or other remote) training and experience to handle the trial? Do they have laptops (with an internal hot spot) that they can lend to jurors for the trial so there is no issue with jurors who may not have internet connectivity (or, poor internet connectivity due to other persons using computers at home, etc.)?

She then lists the "pros" of a remote trial:

1. Jurors do not need to spend wasted time commuting to and from home and courthouse, getting parking, and going through security, which is a savings in time and money for jurors. She stated that she never ran into problems waiting for a juror who was delayed by traffic.
2. There were fewer hardship requests. Elderly jurors or those with medical issues were happy to serve from home with no need to travel to court and expose themselves to possible COVID infection. She also had one juror who said he needed to be out of town for a few days. He was able to remain on the panel by simply logging in from his remote location for the trial.

3. The trials should be conducted in half days because it is unrealistic to have jurors remotely for the entire day. This also results in more potential jurors willing to participate because they have the afternoon to take care of other business or chores they must attend to. She tried cases from 8:30 a.m. to 1:30 p.m. with two 15-minute breaks. Of course, this may result in longer trials given the shorter hours.
4. She noted that exhibits are easier to view and to read on individual screens than in the courtroom.
5 She also believed that witnesses are easier to view closely on a screen, but suspected that attorneys may feel differently.
6. When the need for a discussion among the court and counsel arose, the clerk would place counsel, the judge, and court reporter in a "breakout room," which ensures the privacy of the discussions with no need for whispering at sidebar.

She then listed the "cons":

1. Jurors do not get to socialize or "bond" during the trial—making it arguably more difficult to work with each other during deliberations.
2. Attorneys and witnesses are unable to "connect" with jurors in a personal and visceral way in a remote proceeding compared to being in the courtroom. Not being able to view the entire body and body movements of a witness arguably lessens the ability to evaluate witnesses and their testimony.
3. Monitoring the activities and conduct of jurors is extremely difficult. Jurors are instructed to conduct themselves as if in the courtroom but there is no real way of knowing that they are doing so.
4. Delays and interruptions will likely occur. When a juror is knocked off the internet, the court must respond immediately. If some portion of testimony or evidence has been taken while that juror was off the internet, the court reporter must read back that testimony. Thus, some critical testimony may be heard twice by other jurors, which could be an issue on appeal.
5. The trials are longer due to inevitable delays due to internet issues or due to shorter trial days.

Hybrid Trials

Hybrid trials are those where parts of the trial are in person and others virtual. A number of courts are now using the virtual platform for jury selection and then bringing the jurors in to hear the case in the courthouse. Some in-court proceedings can be supplemented by witnesses, parties, and attorneys appearing virtually while others show up in person. Judges with whom we have spoken say that hybrid trials are more complex to arrange than in-person or totally remote trials. However, they are satisfied that hybrid hearings and trials provide due process to the parties, efficiency to the court, and save time and money for everyone.

Conclusion

My experience with the demonstration trial convinced me that remote trials can be used effectively to carry out the mission of the courts: resolving disputes fairly, with civility, and in a timely manner. And the experiences of the judges in Alameda County, California, and King County, Washington, suggest this is true.

I am convinced that remote hearings and trials provide benefits to the courts and participants that must be factored into the future of the justice system. Hearings involving self-represented parties appear to be a way to obtain better participation and provide the parties with a forum to present their case without having to take time off from work, obtain childcare, travel to the courthouse and find parking, and wait around for their case to be heard. The same is true with regard to represented parties who also should gain a break on attorney fees for not having to pay counsel for travel and waiting time. And the attorneys can use the extra time to handle matters for other clients.

Many have suggested that determining witness credibility is more difficult during a remote proceeding. But I am not sure I agree with that concern. Viewing witnesses, and jurors, remotely is more intimate than in the courtroom. Some of the jurors in the demonstration trial opined that viewing the witnesses remotely was better than in the courtroom. And judges who have handled remote trials agree. Using exhibits and demonstrative evidence may be a bit more problematical than in the courtroom, but when they are placed on the screen each individual juror can look at the exhibit at the same time and counsel can use software to emphasize points to be made with the exhibits.

Some attorneys have opined that they need to be with the jurors in person to effectively present their case. Again, I disagree. In my eight years as a trial judge it was my experience that the facts of the case were what the jurors focused on and that led to the end result, no matter the personality or tactics of the lawyers trying the case. It might take more preparation or a different approach, but lawyers will learn to cope.

With the significant backlogs of cases that have developed over the past year, courts will need to address how to best begin digging out of the hole. I believe remote hearings and trials can provide a way to address the situation. In order to best represent a client and obtain a timely resolution of the client's dispute, attorneys will need to consider using remote trials and hearings when available no matter the nature of the case.[7] Courts may require that certain types of cases be handled remotely, given the authority to do so.

7. *See* Mitchell A. Chester, *Confronting Mounting Case Backlogs Using Creative Strategies and Virtual Jury Trial Technology*, A.B.A. J. Tech eNewsl. (American Bar Association), June 30, 2021.

Chapter 3

Views from Counsel Table

Lisa L. Oberg and Michael Sandgren

Introduction

When the first COVID lockdown orders went into place in March of 2020, no one knew what to expect as to when the courts would be up and running again. In mid-2020, a few courts decided to explore conducting trials remotely to avoid the risks inherent in requiring jurors and the participants in civil trials to gather and spend long days in trial courtrooms. Initially there was great resistance among lawyers to conducting remote civil trials. In a poll of 2,800 lawyers taken by the Texas Office of Court Administration in August 2020, 57 percent of lawyers responded that they would not be open to remote trials.[1] Discussion of the specific objections to virtual trials can be found elsewhere in this book at Chapter 7, Addressing the Legal Permissibility of Online Litigation.

In certain courts, objections based on constitutional, procedural, and technical grounds were overruled, and it was determined that remote trials would proceed. Counsel involved in those trials were plunged, without advance warning, into the utterly uncharted territory of staging an entire jury trial remotely and were required to adapt almost overnight to a completely new platform and manner of presenting their clients' cases. As time has gone on in these courts, the attraction of virtual trials in providing the parties with a manner to have their disputes resolved promptly despite closure of the traditional court system appears to have largely overcome procedural reservations. It seems clear that remote proceedings are likely to continue to be employed at least in some fashion even when they are no longer mandated in order to protect the health and

1. Angela Morris, *Virtual Jury Trial? No, Thank You, Lawyers Say in Court Survey*, TEX. LAW., Aug. 20, 2020, https://www.law.com/texaslawyer/2020/08/20/virtual-jury-trial-no-thank-you-lawyers-say-in-court-survey.

safety of all participants. There is sure to be greater tolerance for the use of remote technology for presentation of witnesses who are far from the courthouse, or who are too ill to easily travel.

Long before actual trial dates approach, counsel should make a point of determining whether remote trials are being employed in the courts in which they practice, and in order to provide the best representation for their clients, the lawyers should be fully versed in the unique aspects of such trials well before having to answer "ready" for trial. The more common challenges counsel may face based on the authors' experiences with remote trials conducted in Alameda County, California, starting just months after the COVID-19 lockdown in June 2020, are explored in the balance of this chapter.

Selection and Control of Trial Technology

The use of video and computer technology in the courtroom is nothing new. As a nation that relishes courtroom drama in real life and fiction, the modern-day juror eagerly anticipates a trial experience that matches their "Law & Order" expectations. Even before the first virtual trial, jurors were routinely shown video testimony, exhibits blown up for all to see, computer simulations, and other demonstrative evidence. As seen in many standard jury instructions, such as California Civil Jury Instruction 5020, counsel and the court have long since embraced the use of technology to "help explain" evidence to a jury and to keep them engaged throughout the trial process.

While technology was already an important component of trial strategy before the COVID crisis, the importance of technology in a virtual trial is absolutely critical. Jurors might excuse an occasional technical glitch, but counsel well know they risk their credibility, and their client's case, with repeated technical failures.

At the outset of any virtual trial, counsel first needs to establish what video trial platform will be used by the court. Has the judge conducted a virtual trial before? Is there a set of established procedures that the court employs? If so, those procedures should be requested and reviewed early in the process, and any ambiguities or challenges should be brought to the court's attention as soon as possible. Does the court plan on using Zoom, BlueJeans, or another video service provider? Does the court have personnel capable of handling the virtual trial platform? How will the court determine whether the jurors themselves have technology that is sufficient to provide them with access to viewing and hearing the proceedings?

Will counsel be called upon to provide them with such technology should the court not have those resources? Will those be recoverable costs for the prevailing party? These are just a few of the appropriate inquiries to the court in order to fully vet the scenario that will greet counsel when they appear for trial. Although the court may be reluctant to reveal any internal deficiencies, counsel will need to candidly explore these issues with the court well before the first juror arrives for voir dire so that solutions can be promptly identified and implemented.

Once a video trial platform has been selected, counsel will need to confirm who operates the platform. In an ideal world, to preserve impartiality, court staff should operate the video trial platform and control the introduction of witnesses, the use of breakout rooms, and other similar functions. However, if court staff do not have the skill level to fully operate the platform, the attorneys should seek a neutral third party to handle these functions. Indeed, even if the court's own staff has primary control over operation of the technology, the use of a consultant to assist with technology at trial is strongly recommended if the budget will allow. Adding the operation of the equipment to the many tasks that face a trial lawyer can be, at best, an added source of stress or, at worst, an invitation to disaster.

As Zoom, BlueJeans, and other platforms have enhanced video and audio control functions, including the ability to mute the participants including the lawyers, a neutral operator will lessen the risk that an unintended technical glitch becomes grounds for mistrial. Since generally only one operator can have rights to the technology at any given time, counsel should always be aware of who has control over simple functions like which participants are placed on "mute," lest counsel find, as one of the authors of this chapter did, that objections are not being heard or recorded because opposing counsel had placed them on mute to attempt to lessen an echo. Even after agreement on a video trial platform and neutral operator, the attorneys and their trial team members must still learn to operate that particular platform competently to avoid trial delays and potentially embarrassing gaffes. It bears repeating that, with some rare exceptions, lawyers are generally best served staying within their areas of expertise and leaving the operation of technology to those who really know what they are doing.

Jury Selection

From the trial lawyer's perspective, jury selection in a remote trial can be absolutely overwhelming. For specific best practices on remote jury selection, referring to Chapter 15 of this book on jury selection is advised. Counsel face unique challenges in conducting jury selection remotely. It is difficult enough to keep straight which jurors have expressed particular views and which should be challenged in a live courtroom setting where they are sitting in a defined space. In the online environment, the most that counsel can see is likely to be juror heads or perhaps torsos in a shrunken "Hollywood Squares" setup.

As a practical matter, it is almost impossible for one person to take in all of the jurors' reactions on the screen at the same time while questioning them, as well as observe the body language of their neighbors. Consideration of retention of a jury consultant is strongly suggested, as the ability to observe and form recommendations on challenges and strikes by someone with experience working remotely can be invaluable. Whether a consultant is used or not, counsel should consider a team approach. While lead counsel examines the prospective juror in the hot seat, other team members can simultaneously view the remaining jurors to observe their reactions and demeanor. By utilizing multiple team members to view prospective jurors, the trial team can collectively analyze juror responses and can formulate educated decisions on which jurors to challenge.

Even in a remote trial, the court will press counsel to move expeditiously through the jury selection process and will be reluctant to delay trial. As it is not possible for team members to "huddle" during jury selection in a virtual trial, an instant messaging system can be utilized to facilitate quick communication. If jury selection is moving very quickly, designating someone as a scribe to record the team's collective impressions of the panel can be helpful as well.

Court Reporter

Counsel needs to make sure that the court reporter can hear them at all times. Even more than in a live courtroom, in a remote setting, it is easy to forget that the court reporter needs to be able to make a clean record since that person will be off-screen and not visible. Without being stilted, counsel must speak even more slowly and clearly into the microphone. And, where possible, counsel should use LiveNote, Realtime, or similar programs to monitor the court reporter's participation. Even the most computer-savvy court reporter can get lost in the Ethernet while trial continues. By closely monitoring the court reporter, counsel can avoid the potential risk of overemphasizing and repeating testimony missed by the court reporter, but not the jurors. This is not a setting where decorum can overcome the imperative of immediately bringing a problem with technology interfering with the record to the court's attention and pausing the proceedings until it can be resolved.

Jury Attention

The art and science of maintaining a juror's attention is not new. But the challenge is much greater in remote trials in which jurors are participating on Zoom or other platforms from their homes. How is a trial lawyer to keep the jury's attention when they are in a familiar and comfortable place (possibly even on their couch or in their bed), wearing comfortable clothes, surrounded by their favorite distractions, and using a device they might normally use to unwind at the end of the day?

Counsel have to demonstrate technical proficiency as a minimum baseline for maintaining jurors' interest in a remote proceeding. Jurors have very high expectations when it comes to the operation of technology, if only because people spend so much time on laptops, tablets, and handheld devices every day, and video meetings have become the standard way of doing business. Fumbling around with the technology ultimately results in jurors becoming bored, irritated, or wondering about counsel's competency, none of which are conducive to delivering a polished and cohesive presentation of a party's case.

It is essential to practice with the technology until it becomes second nature so that it does not stand as an impediment, but instead allows counsel to take advantage of what it can add to a presentation. There are many ways that a comfort level with technology can aid in not distracting the jury with technical glitches and instead keep them paying attention to the case. One rookie mistake can occur in the way counsel handles hard copies of documents. The rustling of paper can make much more noise coming through speakers than expected. (Pro tip: Use a tablecloth or towel and spread documents out on it to minimize the surprisingly loud sound of moving papers around on a hard surface. This is important not least because the sound of crackling paper can take over the microphone, keeping counsel from being heard.)

Lawyers should understand that not only the volume but also the intensity of speech on camera is very different from speaking live. The manner in which the attorneys speak, including while interposing objections, may need to be more modulated in a remote setting as a result.

The difference between a live trial and a remote one can be understood as the difference between being an actor in a live play where it is necessary to project voices and movements to reach audience members in the back rows of the theater and being an actor on television in a close-up shot. Adjustment in tone and volume level should be considered, not to the point of not being heard and understood, but definitely restrained enough to keep the jurors from being put off because counsel sounds like they are shouting, or appears to be reacting with inappropriate facial expressions or gestures. As noted by Judge Brad Seligman of the Alameda County Superior Court, "As Marshall McLuhan once said, this is a cool medium, not a hot medium. It benefits a lawyer to be more conversational than the stem-winder, old-style fire and brimstone type of oratory, because you are literally in the jurors' face."[2]

Often, when lawyers, particularly those without a lot of trial experience or specific performance training, think about bringing their volume down, they also bring their energy and commitment down in equal measure. Sometimes focusing on the distance between the speaker and the audience (in this case the screen) can help the presenter to modulate appropriately without becoming a shrinking violet.

Just as with witnesses who are presented remotely, it is extremely helpful to record and review counsel's performance on screen before trial. How does one's voice come across? How do gestures read in that setting? Is the lighting in the room appropriate and does it show enough, without weird backlit effects or leaving the attorney in the dark, to enable the jury to see the attorney's face at all times? Counsel should consider what is in the room behind them, including whether it is so cluttered that it is distracting, or if there is too much in the way of busy furniture or décor leading the jury to believe that counsel is disorganized and thus possibly ineffective. Avoid virtual backdrops, which with the slightest movement can cause strange lighting effects to make it appear that an attorney's head or body disappears into the background without warning.

Counsel should learn to use the camera so that they can, insofar as it is possible when they cannot be in the same room, connect with the jurors. This is achieved by looking directly at the camera, which can initially feel strange but should be practiced until it can be done naturally and without interrupting the flow of thought or the presentation. Use cue cards that are placed in a way that makes it look like the lawyer is looking directly at the camera instead of looking down at the corner of the screen or even off-screen, causing viewers to be distracted wondering

2. Ken Broda-Bahm, *Don't Be Too Hot for Zoom*, PERSUASIVE LITIGATOR, NOV. 23, 2020, http://www.persuasivelitigator.com/2020/11/dont-be-too-hot-for-Zoom .html.

what counsel is viewing. If the lawyers are comfortable in front of the camera, it will increase the jury's ability to process the information being provided to them, and not be distracted by obvious nerves of the attorney reflected in close quarters over the remote platform being used.

Ultimately, the lawyers' ability to connect with the jury remains as important in virtual trials as in traditional jury trials. As trial proceeds, the entire trial team can assist in observing jurors' responses and engagement. Using the group approach, the trial team can closely monitor individual jurors to ensure that they follow and understand the evidence presented, and that none of the jurors get lost in the process, including falling asleep or engaging in activities around the house.

Trial Exhibits and Evidence

Even in a traditional civil trial, submitting documentary evidence to the court in logical order and confronting witnesses with impeachment materials requires forethought and planning. But, in a virtual trial, these complex tasks become even more critical and important. There are no established court rules or legislation to guide counsel on this subject, so these issues must be handled on a case-by-case basis through the trial. See Chapter 17 of this book for best practices on the handling of evidence.

Determining ahead of time how to handle exhibits such that a witness can view an exhibit and still be seen on the screen by the jurors for credibility purposes is the first step in this process. Even before the first witness appears, counsel must not only ensure that the witness can be heard and seen by the jury, but also must confirm that the witness can receive "trial" exhibits or, for cross-examination, "impeachment" exhibits. (Pro tip: Each remote witness should have two screens (desktop, laptop or tablet) for a virtual trial, one with an integrated camera to facilitate the witness appearance before the jury, and a second for review of the exhibits.)

Once the parties have agreed on the equipment necessary for the virtual witnesses, the next step in this process is determining how to exchange exhibits during trial. By cooperative agreement, the parties can select a document sharing platform to transmit documents to witnesses, significantly streamlining the process. Even in a virtual trial, the traditional rules of evidence apply. Before any trial or impeachment exhibits can be shown to the virtual jury, the court and counsel must still exchange the exhibits, argue over admissibility, and determine whether the exhibits can be displayed to the witness and ultimately the jury. Even in a traditional trial, the admission of exhibits is a slow-moving and complex process. But, if planned poorly, the exchange of exhibits in a virtual trial can add hours to the trial process and can alienate even the most accommodating jury panel.

Counsel should avoid transmitting trial and impeachment exhibits by email at all costs. If an attorney seeks to create a dramatic moment with a smoking gun document or an engaging demonstrative, the instantaneous exchange of documents is needed to guarantee that the entire proceeding does not grind to a halt. If the documents are exchanged by email, opposing counsel and the judge have to wait for the document to show up in their in-boxes, open and review the documents, and address and resolve legal objections. This is an unwieldy enough procedure live, but it is made all the more awkward when performed remotely. The delays inherent in using this method to get documents to witnesses will surely annoy, if not alienate, jurors. Moreover, as the exchange of documents by email in a live trial will inevitably cause lengthy delay, the use of email to exchange documents may, *and likely will*, result in the premature disclosure of counsel's work product. One method that can be used is to have impeachment documents provided to the witness in a sealed envelope not to be opened until directed to by the court, with a similarly sealed set of documents provided to opposing counsel who must openly, on camera, open the package for the first time when given permission to by the court, and remove only those documents that are the subject of the impeachment testimony.

After resolving the necessary equipment and document platform to use at virtual trial, counsel will need to use the available tools to present their best case to the jury. When showing individual documents, the trial team should take advantage of the split screen options available on Zoom, BlueJeans, and similar virtual platforms to effectively display evidence before the jury. (Pro tip: Counsel should practice using the "share" function with their trial team to make sure that documents are easily visible to the jury. Additionally, counsel should sanitize their computer desktop screens to avoid inadvertently "sharing" improper materials during trial.)

While most virtual trial programs do an adequate job of displaying individual documents, significant problems arise with the display of video materials such as deposition testimony or demonstrative recreations. A slow display of a single shared document might not be an issue, but the slow display of video will almost inevitably fail. The jury and court reporter will struggle to hear and understand the halting video presentation. Rather than using the share function to display video to the jury, streaming the video from a separate computer that has the dedicated function of displaying video is a better route. (Pro tip: With court permission, trial lawyers should consider the use of an information technology (IT) professional accustomed to the trial environment to assist so that counsel can concentrate on the content of what is being presented rather than the vehicle by which it is being displayed.)

Opening Statements/Closing Argument

Trial attorneys should consider ways to distinguish the delivery of opening statements and, especially, closing arguments, from the rest of the remote trial proceedings. This signals to the jury that it is important they pay attention. Counsel should consider delivering a closing argument in a different room than they have been occupying for most of the trial. If counsel has been sitting before the camera, consider standing and using a screen with slides displayed on it behind or to the side of the speaker, allowing emphasis on key points on specific slides or demonstratives. Even in virtual trials, the jury still expects an engaging presentation by the lawyers in closing argument. By presenting a new visual display for closing argument, counsel will both satisfy the jury's expectations for a "show" while presenting an effective and memorable closing.

Handled correctly, counsel's use of technology in a virtual trial can impress the targeted audience, whether jury or judge, but this can only be achieved if the lawyer is completely familiar with the technology in use and remains calm when the inevitable glitches in technology arise.

Jury Deliberations

At the close of any jury trial, the jury finally comes together to discuss the evidence presented, the claims alleged, and their analysis. In a traditional in-person trial, the jury is closely observed by the court and provided controlled access to documentary evidence. In a virtual trial, however, the jury cannot meet collectively in a single room and cannot be closely monitored by the court.

Because of the unique characteristics of a virtual trial, the jury may receive a trial binder from the court that includes copies of jury instructions, admitted evidence, and/or a verdict form. While the court might admonish the jurors to withhold review of the trial binders until the jury is engaged in deliberations, some members of the jury might struggle with those instructions. Past experience in virtual trials has shown that human nature is such that jurors will have a hard time not prematurely exploring the contents of such trial binders, convinced that the mysteries of trial are buried within. Counsel should make this point with the trial judge ahead of time, and request that the jury be forcefully and repeatedly reminded to wait until they all come together before reviewing the evidence, instructions, and the verdict form.

As a result of these unique factors, counsel must thoughtfully evaluate what evidence will be admitted and presented to the jury in a virtual trial. And, most importantly, if only part of a story can be told by documentary evidence, it falls on the lawyers' shoulders to make sure the jury understands the full picture. Although the trial may be virtual, the verdict is most certainly not. For additional points on deliberations, reference is made to Chapter 19 of this book.

Observations of Appellate Counsel

Appellate arguments were some of the court proceedings that were the first to go online once lockdown due to the pandemic went into effect in March of 2020. Based on interviews with appellate practitioners, lawyers presenting oral arguments to appellate panels generally agree that virtual arguments can be effective and efficient, saving both time and money for both clients and law firms. It was agreed that virtual arguments tend to feel a bit more relaxed and are certainly less stressful when the lawyer is not required to arrange travel and overnight accommodations for an argument that may last no longer than 15 or 20 minutes. It is also a plus that the lawyers can have all of their materials at hand for easy reference, if necessary, rather than back at counsel table. With most appellate courts, the arguments are livestreamed so clients and others who are interested can observe without having to be physically present at the courthouse.

Questioning from the justices can be a bit problematic, as there can be an awkward switching of screens that has to occur when a justice decides to ask a question. A more productive exchange can be achieved if a gallery view is used so that all participants—counsel and all justices—can be seen simultaneously so counsel can immediately recognize which justice is asking the question. It is very important to remember at all times to exercise the same professionalism as counsel would if they were in a live courtroom. Even when not presenting an argument, counsel should assume that they can be seen and not lose focus and become detached from the hearing.

As in a trial setting, the proper equipment and backup is necessary as well as advance training in the technology. There are differing views on whether counsel should present sitting down or standing at a desk or lectern. Regardless of whether sitting or standing, the background should be neutral.

All agree that one critical element that is missed in remote appearances is the personal touch, whether being able to see the justices face-to-face and observe their body language, or interacting with opposing counsel before or after the hearing. There appears to be consensus that appellate counsel would like to have the option to appear in person or remotely depending on the case at hand and do not have a single preference applicable to every case.

Counsel's Perspective

From the position of a trial attorney, the challenges of remote trials are initially daunting, but certainly can be overcome. Like any other new area faced by trial counsel, adequate preparation can minimize the inevitable challenges that will pop up during a remote trial. Beyond the substantive subject matter of the trial, the added areas that require preparation the most are facility with the technology and careful attention to presentation. Though physically distant from the target audience, lawyers must invest even more energy and preparation into a virtual trial, not least to make sure that their evidence is clearly presented, and their arguments are understood.

Chapter 4

Views from the Witness Box

Alicia Aquino

> *"Knock knock."*
> *"Ms. Patterson?"*
> *"Yes?"*
> *"Here is a trial subpoena for you to personally appear at the Williams Trial."*

This is what typically transpires when a witness is being personally served a subpoena at their residence or place of employment to appear at trial. Once the witness receives the subpoena, they have to make the proper accommodations. They may need to arrange childcare, take the day off from work, borrow a car, or figure out transportation. The witness will have to fight through traffic to get to the courthouse, stand in the long line of security, and then patiently wait in the hallways until the bailiff calls the witness into the courtroom. Often, the witness will wait hours until invited into the courtroom. And in return, the witness is paid a small fee for their time, which in some counties barely covers the cost of parking. Prior to the coronavirus pandemic, this is how the witness would attend trial.

Online Courts Remove Geographic Limits for Witnesses

In some states, a witness can only be compelled to testify at trial if they reside within a certain radius of the courthouse. For example, in California, a person may be compelled to travel only if the courthouse is within 150 miles of their residence. In criminal matters, if the distance is 150 miles or more, but still within California, the attorney seeking the witness's presence in court must present an affidavit or declaration under California Code of Civil Procedure section 2015.5 to a judge explaining why the witness's testimony or evidence is material and their attendance is necessary in court.[1]

Federal Rule of Civil Procedure 45(c)(1) specifically states that a witness can be compelled to testify if they reside, are employed, or regularly transact business in person within 100 miles of the courthouse.

Online Trials Aid Witness Convenience

Generally speaking, online trials are more convenient for witnesses. The witnesses do not have to take a full day off from work. They do not have the hassle of traveling to the courthouse, finding parking, and then experiencing the unknown amount of time spent in the hallways waiting to be called.

In the new norm of remote or hybrid hearings and trials, the witness is provided a link to access the virtual courtroom, and they can easily attend from their home, work, or anywhere there is an internet connection.

The Virtual Waiting Room

As with an in-person trial, there is sometimes a delay of when the attorney *anticipates* the witness testifying for trial, and when the witness actually gets on the stand. Because the remote witness has the flexibility of attending trial virtually, the anticipated time for calling that witness may be more realistic. They will not need to wait around in the hallway. In a remote trial, the witness will wait in the "waiting room" until the judge calls the next witness. The host of the meeting has the ability to send messages to anyone in the waiting room, which is an advantage in easily keeping the witness informed of the trial status. For example, the clerk as meeting host might send the message, "We are wrapping up another witness and will be with you momentarily." If a witness is in the waiting room, and it is unclear who that is, the host can also send a message, "Please rename yourself in your video window identifying your first and last name." This will help the bailiff or judge to ensure it is the correct witness coming into the courtroom.

1. *See* CAL. PENAL CODE § 1330 (2017).

Online Testimony Requires a Technology Check

Prior to the testimony, a technology check should be completed with all witnesses. This includes testing internet speed, microphone, and camera. During this check, the technician should ensure the witness understands how to mute themselves, how to turn their camera on and off, how to position themselves in the camera frame (centered in the screen, with the camera at eye level with the camera), and with nothing that would be either distracting or prejudicial in the camera's background. We also suggest testing the chat feature, raising a hand and screen share features, and any document markup options if those are available. Counsel should prepare the witness for setup and test their technological familiarity in advance. Chapter 13 of this book provides more detailed recommendations on preparing the "studio" for the witness, and Chapter 9 includes advice for that witness preparation session. The final check by the court technician should be a double-check just to see if any troubleshooting is necessary.

In preparing this chapter, we had the opportunity to interview several fact and expert witnesses who have participated in online trials, and found that witnesses appreciate the efficiency of the virtual courtroom as well as the technology check: "Although I had completed a tech check the previous day and my microphone was working great, when it was my time to testify, my mic stopped working. Luckily, I had an extra set of headphones, as suggested by the trial tech assisting the trial team."[2]

Online Testimony Requires Adaptations in Communication

Ultimately, the witness's role is to answer the questions posed by counsel, and the jury or judge will determine the extent to which the witness is credible and provides useful information. That part does not change whether the trial is in person or online. When, however, the witness is testifying in front of a webcam rather than in open court, there are some considerations that are essential to ensuring that the testimony is both credible and effective:

- **Speak slowly, and pause**. This is good advice for any witness, but web-conferencing technology makes it essential advice for

2. Interview with Kevin S., Defendant in a three-week remote bench trial in Orange County Superior Court (Apr. 2021).

the remote witness. Because most web-conferencing platforms cannot handle overtalk, there has to be silence between the end of a question and the start of an answer. Witnesses need to resist the temptation to jump in once they know what they are going to say. The attorney also needs space to lodge a potential objection.

- **Pay attention to credible appearance**. A witness who is well-lit and appropriately framed on camera will do better than a witness who is not. All of the advice in Chapter 13—an appropriate background, well-placed lighting, an eye-level camera, upper body framing—is particularly important for the witness who, after all, is being evaluated based on their credibility.
- **Look at the camera**. Generally, we prefer to listen to and to trust people who are looking us in the eye. For that reason, witnesses should practice looking, not at the screen generally, but at the camera specifically.

Online Testimony Requires Special Considerations in Evidence Handling

In most online trials, evidence is being displayed through the "share screen" feature of the platform. The screen is being shared by either the attorney, someone from the law firm, or a skilled trial technician. When the attorney is examining the witness, he or she will ask for the exhibit to be displayed on everyone's screen, then proceed with questioning. If the exhibit is a multipage document, the witness can ask the technician to scroll through the document and enlarge a certain part of the text if possible.

When we asked the witnesses their perception of the presentation of evidence, they commented that they could see the evidence much more clearly than they would have been able to see it if they were in an in-person courtroom. They appreciated the fact the evidence was ten to 12 inches away from them and they did not have to fumble with locating the document in a specific binder and/or locate the proper tab. It was helpful to see the exhibit on the screen within seconds, allowing for an expedited process.

If an impeachment document is used, or if authentication is required, there are a couple of ways to display the document to the witness without showing it to the jury. First, it can be emailed to the witness to view on their screen. The second way, when impeaching in front of a jury, is to have the witness, judge, attorneys, and court reporter placed in a breakout room, so the document can be shared electronically on the screen to only those who should see it. Once objections are dealt with and permission to publish is granted, everyone can exit the breakout room, and then publish the document in front of the jury. The final way to display the document is to include a link to the document in a chat message directed to the witness, so they can view the document. In some courts, however, the chat feature is disabled.

The overall efficiency of evidence display is a factor that matters to witnesses. One witness we interviewed[3] commented on the display of evidence that "it was clear the same person was not presenting the exhibits on the screen for both parties: one side was definitely more efficient than the other." In this particular case, one party had a trial tech presenting the evidence using trial presentation software, with dual monitors, while the other party had their paralegal assisting with the presentation of the evidence, using Adobe and only one monitor. The paralegal often experienced delays in finding the exhibit and lacked the same features to zoom in, call out, and highlight the document. It was also a distraction every time the paralegal did a screen share; we could see her desktop background, which was an image of Hello Kitty.

Online Testimony Saves Money

Most expert witnesses bill their client in four- or eight-hour blocks, and generally pass along their travel costs, including airfare, mileage, hotel, and meals. With the convenience of an online trial, the expert no longer carries the hassle and expense of travel. If the testimony is only an hour or two, it is also likely they are billing smaller blocks of time because they can attend to other matters during the day.

Online Testimony Adds Comfort and Convenience

The remote setting allows the witness to be in a more comfortable environment. Appearing in court in front of the person who assaulted you, the employer who caused harm to you, or the person who is accusing you can cause stress and anxiety to the witness. It can be less confrontational and more calming for the witness if they are appearing remotely while in a safe environment.

3. Interview with Dennis P., Witness in San Diego family law proceeding (Feb. 2021).

Since the pandemic, the witnesses we interviewed who participated in remote bench trials and mock trials generally reported they felt safer in a virtual courtroom rather than in a traditional in-person courtroom.

Conclusion

There is no doubt in-person trials are different than remote or online trials. There have been hundreds of successful remote or hybrid bench and jury trials since the pandemic. The technical glitches may be unavoidable, but we should remember that there are technical glitches in in-person trials as well. Based on the feasibility and some of the advantages, remote testimony at some level is likely here to stay.

Chapter 5

Views from the Jury Box

Ken Broda-Bahm

As the finders of fact, the jurors are the centers of attention in a jury trial. In an online setting, their overall experience, including their comfort and ability to understand and to manage the constraints of the medium, is critical. The ultimate product of the trial—the jury's verdict—depends on the jurors' ability to be attentive and to follow the communication and the process as it plays out during the online trial, and to effectively discuss and decide the case during deliberations.

Some jurors, at this point, will have prior experience in online meetings using Zoom and other platforms as part of their work and social lives. Some jurors will not have that experience, however. On average, the jury is likely to have the lowest level of technical experience, when compared to the attorneys and the courtroom personnel.

As of press time for this book, we have seen a good number of online trials and experimental projects that have been conducted. At the conclusion of some of these trials and exercises, jurors have been asked about the experience. The common denominator in their feedback is that the experience has been broadly positive, but jurors have leavened that feedback with constructive ideas on how the design and practical execution of the online trial might be revised in order to improve the experience for jurors. Ultimately, the feedback supports the feasibility of presenting a trial to the jury in an online setting while keeping the foundations of the jury's experience intact. As Chief Judge Jack Tuter of Florida's Broward County noted,[1] "Had a judge brought up this possibility of viable jury trials to the legal community in March 2020, it is unlikely anyone would have thought it possible." But after a July 2020 experience with Broward County's first online jury selection, he said, "There is no doubt in my mind jury trials can be conducted via a video platform."

1. Jack Tuter, *Remote Jury Selection During a Pandemic*, BROWARD COUNTY B. ASS'N, July 17, 2020, https://www.browardbar.org/remote-jury-selection-during-a-pandemic/.

Drawing from reports and interviews focusing on jurors' experiences in a number of different recent encounters with online trials, this chapter will identify the most common themes in the jurors' feedback.

Juror Support for Online Trials

The main supportive themes from jurors have focused on the overall effectiveness of online trials in coming close to the in-person experience of a trial, while also noting a few areas where the online format carries a potential advantage.

Broadly, It Works

Jurors note that the online trial experience is able to include all of the functions of an in-person trial and compares favorably to trials conducted in a physical courtroom.

> *I was a juror on a civil case for 10 weeks in 2016. The [online] juror experience was in almost all respects better. I could more easily see the documents, exhibits and witnesses' faces on Zoom than in a courtroom. I could hear everything better as well. Also, I didn't lose an hour each way in traffic every day.*[2]
>
> *This online platform wasn't a negative at all.*[3]

In the Broward County, Florida, online jury selection, all jurors in a follow-up session[4] reported that they could comfortably participate and reported few limitations. As Juror 11 in that group shared, "I am on Zoom all day, every day." She reported that while at first she was nervous, now she is used to the medium. That relatively quick familiarity matches what we have seen in many other contexts, including online mock trials and focus groups.

Ultimately, jurors found the setting to be workable, and for those with prior jury service, familiar:

---

2. Interview with Juror David Lee in Cara Salvatore, *Clear Exhibits, Abrupt Goodbyes: Life as a Zoom Juror*, LAW360, Feb. 9, 2021, https://www.law360.com/whitecollar/articles/1353728/clear-exhibits-abrupt-goodbyes-life-as-a-zoom-juror.

3. Juror 19, post-trial interview in ONLINE COURTROOM PROJECT, ONLINE JURY TRIALS: SUMMARY AND RECOMMENDATIONS (2020), https://www.onlinecourtroom.org/demonstration-report.

4. 17th Judicial Circuit, *Virtual Jury Initiative: Circuit Civil Mock Trial—1st Remote Public Jury Summons—Feedback Session*, YOUTUBE (July 10, 2020) [hereinafter *Virtual Jury Initiative*], https://www.youtube.com/watch?v=VbrhYGmYmQk.

It is clearly still a work in progress . . . but it felt very similar to jury duty in the past.[5]

I thought using Zoom was a very similar experience. I honestly would have to say there was neither an advantage or disadvantage.[6]

The Justice System Is Able to Carry On

In the context of the coronavirus pandemic, it was easy for jurors to see online interaction as a workable solution in a variety of settings, and this "better than nothing" attitude was easily applied to trials as well.

When one juror first heard that his trial would be held over Zoom, he noted: "I was happy that justice was moving forward. I know many cases remain on hold and that's not good for society. I was also happy I wouldn't have to elevate my risk of COVID-19 infection by being forced to be in the courtroom."[7]

At the same time that jurors appreciated the system's ability to move forward, they also noted some differences. In the Broward County online jury selection experience, the judge, Patti Henning, asked, "How many of you felt that you were in court, that you felt the grandeur of the proceeding, the seriousness of the proceeding and what you were doing?" Nearly all jurors raised their hands, but with one unidentified voice chiming in, "Sort of."[8]

There Are Some Efficiency Advantages

At the close of a full online jury trial in King County, Washington, in March of 2021, the Judge, Sean O'Donnell, asked jurors about the experience, and many pointed to the added efficiency:

With respect to voir dire and the ability for us to be in and out so fast, it made that first part super easy . . . I am here for an hour, I'm out. I'm back to work, it was great.[9]

It all seemed pretty seamless to me. Personally, it was amazing just because I live in an area of King County where a major

5. *Id.* (Juror 14).

6. Valerie P. Hans, Virtual Juries (Cornell Law School, Research Paper No. 21-16, 2021).

7. Juror Lee in Salvatore, *supra* note 2.

8. *Virtual Jury Initiative*, *supra* note 4.

9. Interview with King County Superior Court Online Jury Post-Trial (Juror 7) (Mar. 18, 2021) [hereinafter King County Interview].

> *bridge went out at the beginning of COVID . . . I am working from home, I am on Zoom calls, video calls, for work all the time. So that was an easy thing to slip into my calendar, and not have to interrupt the rest of my workday, and without the additional stress of figuring out how to get downtown.*[10]

This potentially translates into a greater willingness to serve on juries generally if jurors are provided with an online option: "If I'm honest 100 percent, I would have felt very differently about serving on a jury if I had to drive downtown and go to the courthouse. I really appreciated being able to do it online."[11]

Being at Home Is More Comfortable

While the home location may not carry the same grandeur as the courtroom, it did have other advantages for many jurors:

> *Most jurors had cats, and at least one had a dog. These pets made occasional appearances. I personally found it great that my cat could hop on my lap from time to time. . . . Being at home for breaks was handy!*[12]
>
> *Virtual jury [means] you don't have to wear pants (kidding). You can have your own home-brewed coffee by your side versus some mock jury coffee or maybe they don't even have coffee during it.*[13]
>
> *As a disabled person, being able to attend to my needs while also being able to attend this event was a huge advantage. I truly hope we as a society start making more use of the advances technology gives us where accessibility is concerned.*[14]

When jurors in the Broward County online selection were asked in the follow-up interview about what they would prefer if given a choice between conducting the trial in person or via a connection in their home, most said they would be more comfortable at home, with a few saying that either would be fine, and none saying they would prefer the physical courtroom.

The most vocal jurors, however, emphasized the convenience of not having to come into court:

10. *Id.* (Juror 9).
11. *Id.* (Juror 10).
12. Juror Lee in Salvatore, *supra* note 2.
13. HANS, *supra* note 6.
14. *Id.*

As a parent, the burden taken off by eliminating the commute makes a lot of difference in the ease of participation and the decision on whether to defer . . . the ease of not having the commute makes a huge difference.[15]

I didn't have to drive anywhere and find parking![16]

The On-Screen Visual Field Has Some Advantages

While the act of distilling all of the sounds and sights of a courtroom down to a single computer monitor might seem to be limiting the visual field for the online trial, many jurors reported that it was actually easier to see. Seeing the attorneys, the judge, the witnesses, and the exhibits on their own monitors provided a close-up view of everything, which is a perspective that is not replicated in the physical courtroom.

I could see all the exhibits very well. I could see them better than at the in-person trial discussed above. Once the jury went to deliberations, documents were available on box.com. That experience was much better than an in-person trial where all jurors must crowd around a single copy of the documents.[17]

One advantage was that all of us saw the same thing all at once, rather than having to pass things around.[18]

It was really nice being able to see the other jurors face-to-face. Usually, we would be sitting side-by-side [during voir dire and trial], not able to look at each other.[19]

I feel that over Zoom we are a closer group. Closer physically. Faces up close and personal.[20]

In the Broward County jury selection, none of the jurors indicated in the follow-up interview that they felt they would have been at any disadvantage in judging the credibility of the witnesses based on their view of courtroom participants.

Deliberation Is Still Effective

One of the biggest worries for skeptics contemplating the online jury trial is the moment when the jury is on their own, unable to be monitored,

15. King County Interview, *supra* note 9 (Juror 3).
16. Hans, *supra* note 6.
17. Juror Lee in Salvatore, *supra* note 2.
18. King County Interview, *supra* note 9 (Juror 13).
19. Online Courtroom Project, *supra* note 3, at 40.
20. Hans, *supra* note 6.

and expected to lead their own unaided discussion to a verdict. While complications are certainly possible in this setting, the early experience has been that jurors are able to manage the deliberations on their own. In the Online Courtroom Project demonstration, for example, one of the conclusions is that "[o]verall, jurors felt that the online trial process did not affect their ability to discuss and debate the merits of the evidence as a group."[21] The pacing of deliberations online, as well as the technologically enforced need for there to be just one person talking at any given moment, also creates the possible advantage of jurors being more deliberate when speaking in deliberations. For example: "Zoom did not create any difficulties in deliberations. I've found that people are better about not talking over others on Zoom calls than they can be in person. That held true in this case."[22]

Social Connection (Bonding) Is Still Possible

A feeling of affinity and common purpose is considered important to the ability of a jury to candidly discuss the case and reach a decision. One concern with online trials is the question of whether jurors seeing each other only as faces on a screen will be able to bond as much as they do in person. Despite the initial awkwardness of the online settings, jurors do report an ability to get to know each other via an online connection:

> *The jury was asked to connect to the Zoom call up to thirty minutes prior to resumption of the trial each day. Clerks stated at least once that this was so that jurors could get to know each other. Items behind jurors (pets, posters, guitars, etc.) provided great starting points for conversation. Socialization wasn't as good over Zoom as it was for me in person at the prior trial, but the duration for this trial was much shorter, so that's expected. I feel like socialization was in a good place prior to deliberations.*[23]

Some jurors additionally saw an advantage to the online interactions due to the reality that everyone on a Zoom call appears to be looking at everyone else:

> *I think it's interesting because trials and the whole process of the trial can seem very formal and very impersonal, but because we are facing each other and looking at one another, it seems like there is an extended level of connection that you don't*

21. Online Courtroom Project, *supra* note 3, at 44.
22. Juror Lee in Salvatore, *supra* note 2.
23. *Id.*

experience in the courtroom and you get to know each other in another way whereas in the courtroom you're all sitting face forward and looking straight ahead. So, I think there is a newfound level of connection that you wouldn't find in a brick and mortar courtroom.[24]

For some jurors, the "face and shoulders" framing of the video camera causes a focus on facial expression that seems to be experientially more potent in the online environment than it is in the physical environment: "I think when we got to the jury room, I feel a lot more connection, and even with the witnesses because you are right there. You can really see their expressions."[25]

The Jurors' Perceived Challenges of Online Trials

While juror feedback has been generally positive, those who have gone through an online trial experience have also shared feedback on the difficulties in that format and some of the ways the interface with the jury can be improved.

Attention Can Be More Difficult

As online meetings have become the norm during the pandemic, academics have warned of a "Zoom fatigue"[26] that can set in, making it harder to pay attention for long periods of time. Because they need to sit through the better part of a day keeping their attention on the screen, and possibly due to other factors such as the self-consciousness of seeing their own image on the screen, jurors can become exhausted by the unique attention demands. While Judge O'Donnell in the King County case commented to jurors that he had been watching them throughout trial and "You were all super-engaged," he maintained that appearance of engagement can take its toll. As one juror noted: "One downside of Zoom was that I felt compelled to keep my face pointed at the camera the entire time so it was clear I was paying attention. In a courtroom you can occasionally look around the room and out the windows."[27]

24. Online Courtroom Project, *supra* note 3, at 44 (Juror 13).
25. *Id.* (Juror 19).
26. Jeremy N. Bailenson, *Nonverbal Overload: A Theoretical Argument for the Causes of Zoom Fatigue*, 2(1) Tech. Mind & Behav. (2021).
27. Juror Lee in Salvatore, *supra* note 2.

That difficulty in maintaining a single focus might argue for more varied display settings. For example: "I would have just liked to see everyone at once. I like to see reactions and how people are responding. So rather than just the speaker or just the document, I would rather see everything."[28]

One additional attention challenge is simply the potential distraction in being in one's own home, and the question of whether there are unique distractors in a juror's personal environment: "Being here at home, there are a lot more distractions than there would be in the courtroom. Am I going to pay more attention in the courtroom than I would here?"[29]

Body Language Reads Differently

Part of the complexity of communication is the fact that we attend to each other's words as well as intonation and physical behavior. While we are able to see and hear that during videoconferencing, there is also at least the perception that we are getting a limited or altered view of that nonverbal communication. As one juror notes: "There is much less body language over Zoom than in person."[30]

Importantly, some jurors disagree with that: "You can read a person's body language just as easily on a computer screen as in a courtroom . . . some people might disagree with me, but I don't see a whole lot of difference."[31]

One difficulty is that jurors can be uncertain of where to look on the screen to see the person who is talking:

> *It randomly picks the order to put the video tiles in, and it was annoying that during witness testimony, more often than not, it did not put the witness and the attorney asking questions anywhere close to each other . . . so I could only see one at a time.*[32]
>
> *My head's moving all over the place just to see who is talking.*[33]

There is also the tendency that some have observed that, when we are watching communication on the small screen instead of viewing it live, there is less latitude for emotionally intense communication: within the four corners of the small screen display, emotionality can look

28. King County Interview, *supra* note 9 (Juror 2).
29. *Virtual Jury Initiative*, *supra* note 4 (Juror 6).
30. Juror Lee in Salvatore, *supra* note 2.
31. *Virtual Jury Initiative*, *supra* note 4 (Juror 5).
32. King County Interview, *supra* note 9 (Juror 7).
33. *Id.* (Juror 10).

over-the-top, or "too hot for Zoom."[34] For example, in the King County online trial, one attorney got aggressive with a particular witness and admitted in the post-trial interview with jurors that he was doing that somewhat "performatively" in order to make a point. One juror responded:

> *I see "performative" elements of this being really old school, and it makes me lose trust because I feel like I'm being performed to or entertained in an unnecessary way. . . . We are all humans, and . . . it just seems silly to me. Just making that connection, because we are over Zoom, is a little bit helpful.*[35]

Interaction Is More Challenging

While it may be a function of simple familiarity, some jurors do find it more difficult or more awkward to interact as they normally would in person:

> *You are on mute and you feel like you need to wait to be called on.*[36]
>
> *It is harder to judge the body language of when someone is finishing up . . . those kind of social cues that we're used to are harder, so it is more of a raise-your-hand situation.*[37]

Both of these jurors were talking about interaction during voir dire, but one could easily see this uncertainty applying to deliberations as well.

There Is a Learning Curve

With any novel technology, there will be a "break-in" period of time where people become familiar with it. For online conferencing during the last three quarters of 2020, that period seemed to happen quite rapidly. For many, the Zoom conference is now routine. But that familiarity is not yet uniform, and we can expect some in the jury pool to still be learning. For example, in the state of Illinois's first civil Zoom trial,[38] two jurors

34. Ken Broda-Bahm, *Don't Be Too Hot for Zoom*, PERSUASIVE LITIGATOR, Nov. 23, 2020, https://www.persuasivelitigator.com/2020/11/dont-be-too-hot-for-zoom.html.

35. King County Interview, *supra* note 9 (Juror 9).

36. *Virtual Jury Initiative*, *supra* note 4 (Juror 23).

37. *Id.* (Juror 14).

38. Clifford Ward, *Groundbreaking Civil Jury Trial over Zoom Begins in Lake County, Drawing Interest from Legal Community*, CHI. TRIB., Feb. 18, 2021, https://www.chicagotribune.com/suburbs/lake-county-news-sun/ct-lns-first-virtual-jury-trial-st-0218-20210217-4gyk7wncwna4fi6zec4c3mfk5u-story.html.

reported having technology issues, but were helped by their children. In the Broward County trial, Judge Tuter noted some difficulties there as well:

> *Some jurors expressed difficulty with operating video equipment primarily with the use of iPhones. Some stated they could only view four video panels on Zoom and felt uncomfortable participating; others claimed weak Wi-Fi signals; some made excuses just to bail out such as being at a car repair shop when they checked in.*[39]

It is likely that some of these difficulties can be solved through better communication. For example, Judge Tuter noted that in Broward County's first online juror summons, a majority of those responding did not read the instructions and did not see a link to fill out a pretrial questionnaire, and, as a result, they did not complete it in advance.

Sound and Video Glitches Can Occur

Of course, anyone who has participated in a web-conference call, or used online technology generally, knows that breakdowns can occur. Someone loses audio, someone else's video freezes. In some instances, it is a simple matter of not knowing how to unmute for example, but in other cases, the glitch remains a mystery that is only solved when the computer is rebooted, or the connection is restarted.

Jurors in early online trials have experienced these problems as well. In the Online Courtroom Project's demonstration in the summer of 2020, for example, two jurors discussed the extent to which these gaps reflect badly on the parties trying their cases or not:

> *Juror 13: The only thing about yesterday—and it didn't take away from their professionalism at all—I had this empathy factor any time any of the things froze up, and so that was the only thing that took me out. And I was thinking "I feel so badly for them" because it's their exhibit and they want to show something and they're not able to show it, and it didn't reflect on their professionalism whatsoever.*
>
> *Juror 15: I think actually the exact opposite of the empathy thing. Even though I know they are technical issues and that things happen, for me, it's just subconsciously. If I see someone having technical issues on a platform that I feel like I can*

39. *Virtual Jury Initiative*, *supra* note 4.

> *use—there is just something that's like, "Okay maybe you don't know what you're doing." That's probably just me personally and something I need to fight and work against, but I do think you know if one side is having more technical issues than the other side, there might be a subconscious thing there.*[40]

The most basic sources of problems on a web-conference are often unwanted noise. While the hosting platforms continue to make advances in dealing with this problem, and participants come to understand the importance of everyone other than the speakers muting their microphones, some problems emerge from unexpected sources. For example, in the King County online trial, the culprit turned out to be the judge's chair. One juror explained in the post-verdict interview:

> *It goes with the territory, but if there is any cross-talk at all, or any noise in the background, it cuts off who is talking . . . in particular [to the judge], your chair has a squeak, and you would sit back in that chair and the squeak would cut out three words of a sentence. . . . Those little noises can be a problem.*[41]

In subsequent discussion, the judge and juror decided that the problem could be solved through a more sophisticated microphone, or a can of WD-40.

There Are Delays

Delays are, of course, part of the trial process, and what can seem superficially like wasted time to the jurors, is often devoted to critical issues for the litigants. In some ways, the online format reduces some of these delays—for example, once the clerk is familiar with the way it is done, excusing or recalling the jury takes seconds rather than minutes. However, there seems to be a lower tolerance for sitting through any delays in an online trial setting, and once participants are in front of their screens, they are ready for things to happen.

In the Online Courtroom Project's demonstration, for example, there were some logistical delays in sending the jury to deliberate. Because there was also a second "shadow jury" in that case, there was a mistake regarding which jurors were to be sent to which rooms, and one of the shadow jurors ended up spending substantial time off camera and eventually dropped off the call.

---

40. Online Courtroom Project, *supra* note 3, at 46.
41. King County Interview, *supra* note 9 (Juror 7).

In actual trials, jurors have also commented on delays, but considered them a relatively modest issue:

> *The trial had a handful of small pauses due to typical Zoom patterns including people being muted/unmuted at the wrong time and people not finding someone in the grid of faces. But there were no substantial issues. The trial was paused for a few seconds when a dog jumped into frame, but that was about it.*[42]

Some In-Person Features Cannot Be Easily Replicated

The final theme in jurors' discussion of challenges relates back to the comparison between an online and an in-person trial and the realization that some features of the latter cannot be replicated or can be replicated only with difficulty in an online setting. These differences are sometimes very specific and are realized only when the need comes up. For example, in the King County trial, one juror expressed the wish to have a way to send a note to the bailiff without needing to switch programs by sending a text or an email. In order to discourage irrelevant conversation, Zoom's "chat" feature had been disabled for jurors, but courts may want to consider options to allow a direct line to the bailiff, which would also be useful in letting the court know when the jury has reached a verdict.

Jurors have also reported wanting a way to conduct a secret ballot, as they might have using slips of paper if they were in person: "Also, when you do change your mind anonymously, I think that that is easier than changing it once everybody has seen your vote[. A] lot of people don't want to admit they're changing their minds."[43]

In theory, it should be possible for a platform to use its "polling" function to permit the jury foreperson to execute that sort of poll, revealing the total votes without necessarily revealing how each juror voted.

One other important feature of the trial that is difficult to fully replicate is the jury's waiting room. In an in-person trial, this is the location where the group chats, trades snacks, and starts to get to know each other. That familiarity and common ground is important to the rapport that facilitates a good deliberation. In an online setting, it is possible for jurors to have their own "room," but it is still more difficult and awkward to have natural informal interactions. In the physical jury room, a couple of jurors might be reading novels, while a pair has their own conversation,

42. Juror Lee in Salvatore, *supra* note 2.
43. Online Courtroom Project, *supra* note 3, at 43 (Juror 33).

and a larger group trades jokes. That is likely impossible to replicate in an online space where all of the conversation is for everyone.

One juror pointed to this as the one area where the in-person trial has a significant advantage: "The only point at which it would have been dramatically better to be in the same room was after the trial so that I could say goodbye to each fellow juror. It felt strange to just collectively wave goodbye and leave the meeting."[44]

Conclusion

As online trials become more common, we will want to conduct more comprehensive surveys of the jurors and to more systematically test some of the ways to leverage the online trial's strengths while addressing its challenges. Educational groups, such as bar associations or the Online Courtroom Project, can facilitate this process by encouraging courts to adopt a common questionnaire to be administered at the conclusion of the trial. That broader data collection may point to additional issues.

At present, however, the view from the jury box is that, while there are some real challenges in moving courtroom communications online, there is also a surprising amount of success, with jurors generally saying that they are able to attend to the trial, assess the parties and witnesses, and reach a decision, all from behind their computer screens.

44. Juror Lee in Salvatore, *supra* note 2.

Chapter 6

Meeting the Technical Challenges of Online Litigation

Sarah Murray, Josh Splansky, Marc King, Noah Wick, and Ted Brooks

"It is not the strongest of the species that survives, nor the most intelligent that survives. It is the one that is the most adaptable to change."

—Charles Darwin

Introduction

As anyone who has spent time in courtrooms and courthouses knows, judges and lawyers are not typically early adopters of technology. The law as an institution tends to be conservative by nature, as laws are designed to fix common understandings in place and be hard to change. This is why the law attracts people who are conversative by nature—not necessarily politically conservative, but personally conservative, valuing and appreciating tradition, routines, rules, and regularity. A saying Hon. Richard B. Klein often shares is, "Attorneys and judges are all for progress, but one hundred percent against change."[1] Trial in particular is also a very conservative ritualistic form that has resolutely remained focused on in-person, oral, face-to-face interactions in a world that has moved more and more to mediated meetings and communication. In addition, trial work entails extremely heavy workloads for all of the professionals involved—judges, attorneys, support, and courtroom staff—which

1. Richard B. Klein, *Lawyers Are All for Progress, but 100% Against Change*, TIPS FROM BENCH, Apr. 22, 2020, https://www.linkedin.com/pulse/lawyers-all-progress-100-against-change-richard-b-klein.

makes it difficult for them to learn new technologies and develop new work routines, even for those who seek to innovate and share those innovations with colleagues.

These are some of the reasons that pre-COVID, many courtrooms and many lawyers were only slowly moving from the world of paper documents, in-person meetings, tabletop cameras, and telephone conferences into the world of video evidence and presentations, online questionnaires, graphic presentations, videoconferencing, and full online trials.

Another important reason, however, is that because of the relative lack of technical expertise among judges and lawyers, many in the profession simply could not imagine how it could be done—how trials could be brought online in ways that would work, meet all of the constitutional and practical requirements, be financially feasible, and ensure that justice was still being done.

The coming of COVID-19 forced the issue. The closings of courts that took place starting in March of 2020 did two things: it stopped the daily merry-go-round of court proceedings and trials and gave overworked judges, court administrators, and trial attorneys time; and it pressed judges and courts around the country to find ways to use technology to hold court proceedings, like it or not. With a choice between the justice system being shut down and a justice system adapting technology to serve citizens, all of a sudden, using technology to enable remote proceedings started to look pretty good to a lot of participants in the system. So began experiments in jurisdictions around the country.

Our group, the Online Courtroom Project (OCP), was part of this process of experimentation. As a collection of judges, attorneys, trial consultants, and trial technologists, we saw right away that COVID-19 would be a time of reckoning and profound change for the courts, particularly the trial system. Some of us are professionals who have spent our careers helping attorneys and judges understand how to leverage technology effectively in the courtroom; some of us are consultants, attorneys, and judges who have been early adopters and sought to educate colleagues to new possibilities. All of us are people who care about ensuring access to justice and ensuring the quality of decisions made in the justice system.

The OCP felt that limiting ourselves to debate about whether online trials were a good idea was unproductive. Under conditions of COVID, courts were closing and justice and conflict resolution—fundamental needs of our society—were becoming unavailable. We thought the best way forward was through practice—rolling up our sleeves, going to work, and producing online trials to experience directly what worked, what did not work, and what happened that nobody had considered before. That led us to produce one of the first full online trials in the country, in June 2020. It was a demonstration trial, based on materials from a real personal injury case—a slip and fall case—on which two of our members had previously worked. The idea for a new role, that of the "technical bailiff," emerged from that demonstration trial. That is typically how innovation and learning work—through practice.

Many of the authors of this chapter have since then gone on to produce or provide technical support in dozens of online bench and jury trials, hearings, and other trial proceedings over the past year in public courts and private trials around the country.

We are of course quite aware of the many objections that have been and continue to be raised against online trials. Those objections generally fall into three categories:

- **Technical**. It cannot be done technically, or the technology is too expensive, insufficiently available, or improperly featured to successfully mount online trials.
- **Legal**. Holding trials online violates the rights of the litigants in fundamental ways that cannot be overcome.
- **Substantive**. Online trials fundamentally prevent key tasks of trials, like evaluating witness credibility or presenting evidence, from being carried out properly or being carried out at all.

This chapter is a response to technical objections related to the operations of the courts and trials. It is not a compilation of specific technical tips and tricks; we offer that elsewhere in this volume. Instead, it summarizes what we have learned about how to make technology work, in the present moment, from a high-level perspective, from the perspective of systems and organizational design and social and cultural practices, based on our own practices and what we have learned from judges, court administrators, lawyers, and other trial technicians around the country who have been on the ground, figuring out how to make online proceedings work under COVID conditions.[2]

2. Josh Splansky and Marc King conducted formal interviews with trial technicians Noah Wick and Ted Brooks, members of the OCP and co-authors of this chapter; and trial technician Rob Rosenberg. Between the three of them, over the past year, they have conducted more than 100 online court hearings, bench trials, and jury trials. Splansky and King also interviewed Judge Sean O'Donnell of King County Superior Court (Seattle, Washington); Phil Doreau and Lynne Shannon, information technology (IT) and audio/video (A/V) administrators for the U.S. District Court for the District of Massachusetts; Jim McMillan, principal court technology consultant for the National Center for State Courts; and David Slayton, head of the Texas Office

The good news is that while the idea of online trials may seem new and different, courts around the country have successfully adapted to the point that, in less than a year, remote proceedings are now a daily occurrence—and they are working. By all accounts, they are also making court proceedings more accessible for many litigants and jurors.[3]

This rapid adaptation is not as surprising when we recognize that technology already played an important role in the court system and the courtroom even before COVID-19. For years, judges have used telephone conference call systems to hold remote hearings and allow remote appearances, and many have also used videoconferencing systems. Many courtrooms have built-in projectors and screens to allow electronic presentation of evidence, which has become commonplace, especially in complex civil litigation, which often involves tens of thousands of documents. Most courtrooms have sound systems and microphones to amplify and record sound.

So the type of technology online trials require is not significantly different in kind from the "old fashioned" courtroom trial presentation technology we have been using successfully for the past 20 years in support of in-person trials. The *technical* steps required to move forward are not as big as many have made it seem.

The social, political, and cultural steps required, however, are bigger. That is why what we believe is needed to move forward is collaborative development of process flows that integrate technology into trials and practices to create social trust in the mediated encounters of an online jury selection or trial; and policy work to develop new visions of how to carry out the work of justice with existing resources. We do not necessarily need more technology or more money; we need better thinking about how to use well the technology and money we already have.

In sum: technology is relatively easy. Change and process are hard—but they are readily achievable if we work together and keep our eye on our common goal of providing consistent and accessible justice for all Americans.

---

of Court Administration. We also draw on comments made during presentations for the OCP-National Institute for Trial Advocacy summit on online trials in November 2020 and interviews conducted of judges and David Slayton by Marc King and Sarah Murray for Chapter 8 of this volume.

3. Please see Chapter 8 for a detailed discussion of the representativeness of online jury pools and a review of data on Americans' access to computers and broadband internet access.

How This Chapter Is Organized

We begin our chapter by discussing the state of technology in state and federal courts prior to COVID-19.

We then organize the chapter around a series of objections commonly voiced by those who resist or reject online trials:

- Online trials cost too much
- The unique expenses are too great
- The learning curve is too high
- Managing online jury selection is too hard
- You cannot pick a jury through a screen
- Monitoring juries is too hard
- Document handling is too hard
- Public access is too hard
- Online trials are too insecure
- Online trials cannot reproduce the gravitas and power of in-person trials

We consider, discuss, and—cutting to the punch line—reject each of these objections in turn from the narrow technical point of view, based on our experience and the experience of others who are actually mounting online trials and proceedings.

Simply said, from a technical point of view, online trials can be done successfully because they have been done successfully, over this past year and before. Of course, online trials are not problem-free—just as pre-COVID in-person trials were not problem-free. Courts and trial technicians experienced bobbles and glitches as we moved trials online and we have learned, adjusted, and found technical solutions. We have also reached out to commercial providers of software, built relationships, and started the process of educating providers about the unique technical needs of courts and trial participants.

However, we do see important issues that naturally arise from the technology involved in online trials that courts, attorneys, and the public need to consider and address if online proceedings are to become routine. In other words, online trials can be done but we have not yet found optimal solutions for every challenge or solved every problem. We also anticipate technical challenges that will arise going forward—such as the challenges of storing large quantities of video and creating more online-friendly courtrooms, as we discuss further below.

We also note that people who say that online trials cannot be done or pale in comparison to in-person trials are always referring to in-person trials before COVID. As of our publication date, in-person trials continue to be subject to many limitations and due to social distancing requirements, creating conditions that in many ways undermine or even undo the most important aspects of physical presence. Witnesses testify behind plexiglass tens of feet away from jurors, many of whom now sit in in the gallery rather than the witness box. Some witnesses may even wear masks. Jurors do not mingle in the hallways or jury room and rub shoulders as they did before. Everything we discuss in this chapter presumes pre-COVID trials. As of press time for this book, however, the time frame is uncertain for returning to "normal" courtroom proceedings, and the potential for future disruptions remains with us. Most of these issues require political, social, and cultural solutions rather than technical solutions, or a combination of technical and social solutions, although there are still some low-hanging technical "fruit" to be picked. Court budgets and government priorities need to continue to evolve if online trials are to become a routine offering, and stakeholders in the justice system need to come together to develop best practices that create the best experience possible for all participants. That is why we conclude by sketching areas where we believe innovation, more robust technical solutions, and, most importantly, public dialogue and conversations are still needed.

Courtroom Technology Pre-COVID

While lawyers and courts may have been slow adopters, our court system has been just as impacted by the emergence of desktop computing, the internet and cloud computing as the rest of society. Even before COVID, our courts and trial practice had been transformed by these new technologies.

Innovations that have impacted courtrooms and trials include:

- Desktop computers and word processing, replacing typewriters
- Spreadsheet and statistical software, which has allowed for better record-keeping and data analysis of court activities, as well as the creation of randomly generated jury pools for jury trials
- Email and internet connectivity
- E-filing systems that allow for electronic filing of documents
- PDF documents and the development of secure technologies for digital encryption, digital signatures, and digital watermarking
- Telephonic conferencing systems
- Court reporting technologies for recording and transcribing court proceedings, particularly systems that allow for immediate, live transcripts on computer screens for judge, staff, and lawyers

- Trial presentation software that allows for real-time presentation of evidence through projection and video systems
- Videoconferencing systems
- Projectors and monitors in court

The federal courts were the first to install courtroom presentation and remote appearance technology system-wide. As is typical in our federal system, where individual states run and fund their state court systems, the installation of technology in state courts around the country has varied in timing and quality. You are more likely to find technology in state courts located in busy metropolitan jurisdictions. In rural areas or smaller courts, technology may be available only through portable technology units or private trial technicians on a per trial basis. Regardless, remote appearances by attorneys, parties, and witnesses in criminal arraignments and presentments, foreign language translators, status conferences, and hearings and trials by telephone and video were already routine in both state and federal courts around the country well before COVID-19.[4] Rob Rosenberg, a leading trial technician, commented that, for the courts, providing technology support for online trials was not a big leap because

> a lot of courts are dealing [with] criminal matters and oftentimes they're having to have criminals testify from the jail via video conference. So, a lot of this was already built into the system . . . I didn't feel like the learning curve for courts was overwhelming. Courts generally have a pretty robust staff.[5]

Court budgets and public reports reflect this. In its 2019–2020 budget request, before COVID-19 was even a twinkle in the eye of the world, the New York judiciary wrote that the provision of justice "necessitates the ongoing provision of cutting-edge technology in all aspects of court operations, including electronic filing, remote access to court proceedings, and an ever-expanding list of operational demands."[6] In the same submission, the court reported proudly about

---

4. Herbert B. Dixon Jr., *The Basics of a Technology-Enhanced Courtroom*, 56(4) Judges' J. 36 (2017), https://www.americanbar.org/groups/judicial/publications/judges_journal/2017/fall/basics-technologyenhanced-courtroom/.

5. Interview by Marc King and Josh Splansky with Rob Rosenberg, President, Rosenberg Consulting Services, Inc. (Apr. 20, 2021) [hereinafter Interview with Rob Rosenberg].

6. Janet DiFiore et al., New York State Unified Court System, Budget, Fiscal Year 2019–2020 (2018), http://ww2.nycourts.gov/sites/default/files/document/files/2018-11/2019-20-JUDICIARY-Budget.pdf.

> the statewide expansion of [the state's] remote appearance project, permitting the electronic filing of family offense petitions and ex parte temporary orders of protection by audio-visual means (Skype). The program allows vulnerable crime victims, such as the elderly, disabled, or victims of family violence, to seek relief that was previously difficult to obtain due to lack of transportation, mobility issues, safety concerns or poor health.
>
> To facilitate the presentation of evidence at trials on various technology platforms in a cost-effective manner, the court system has developed an Integrated Courtroom Technology (ICT) program, which combines sound enhancement, assistive listening, audioconferencing, video conferencing and WiFi in a cutting-edge touch screen format that can be employed seamlessly and economically throughout the State. It is anticipated that this powerful technology will be in place in every county and court in the State over the next five years.[7]

"Online Trials Cost Too Much"

We spend a lot of money on our court systems in the United States. In 2017, a tally of state budgets for courts came to $18,088,251,609.[8] (For those of you who are number challenged, that is more than $18 billion.) The states with the five largest budgets were California, New York, Massachusetts, New Jersey, and Colorado. Notably, Texas and Florida, two of the most populous states and two of the states to most quickly adopt online trials, are not among the top five in budgets: Florida is number 11, with a budget of $504,954,746, and Texas is number 13 at $397,957,159.[9]

A significant amount of those costs come from the costs of managing physical courthouses—building and managing buildings, providing security, furnishing offices and courtrooms—and, of course, providing information technology.

All of these state budgets included line items for information technology. Technology in the court system and the courtroom is here already to serve and to stay; the question is not whether to obtain technology, but

---

7. *Id.*

8. Figures come from Ballotpedia, *State Court Budgets and Judicial Salaries, 2012–2017*, https://ballotpedia.org/State_court_budgets_and_judicial_salaries (last visited Nov. 6, 2021). The authors performed computations to arrive at totals.

9. The five most populous states are, in descending order, California, Texas, Florida, New York, and Pennsylvania.

how to prioritize and enable its use in the administration of justice. Still, the questions remain: how much would moving to online trials cost, and do online trials cost too much?

When discussing the costs of online trials, we benefit if we distinguish between three sets of potential costs:

- Costs of providing judges and courtrooms with the needed technology, training, and personnel to broadcast and stream trials; to conduct hearings and other proceedings; and to preserve video recordings
- Costs of providing technology and training to litigants who are being asked to appear in online proceedings but may not have the needed equipment to participate
- Costs of providing technology and training to jurors who are serving in online jury trials, to ensure that all jurors have access to appropriate technology to allow full participation

For ease of reference, we will refer to the first set of costs as "court technology costs," the second set of costs as "party technology costs," and the third as "juror technology costs." It is helpful to distinguish between them because the management challenges and potential funding sources are different.

Court Technology Costs

Most federal and state court systems in the country already have in place a robust suite of technology, including high-speed internet service; cloud data storage; e-discovery and online filing systems; and courtroom presentation technology, including digital projectors and monitors in the courtroom. For that reason, the costs of moving to online trials are not as much as many critics suggest. Yes, moving to online proceedings and trials has required all the courts of which we are aware to expand their software licenses and obtain new equipment of some kind. For example, King County quickly moved to expand their number of Zoom licenses once it became clear that there would be no "business as usual" in the King County courts for some time. The state of Texas procured 3,000 Zoom licenses, one for every judge in the state, and began training the judges right away.[10] However, since courts are already using technology and will continue to do so whether or not online trials are conducted, enabling courtrooms and judges to offer online trials as an offering in

10. Arya Sundaram, *How Texas Courts Went Virtual*, TEX. OBSERVER, Feb. 10, 2021, https://www.texasobserver.org/how-texas-courts-went-virtual/.

most cases entails rethinking or reallocating existing resources rather than funding new ones.

In addition, it is important to consider:

- What kinds of cost *savings* might also be entailed in moving some or all phases of trial online?
- What are the costs, financial and social, of *failing* to offer online proceedings as an option, now during the COVID-19 pandemic and going forward, even when the pandemic is under control and in-person trials can resume?
- How might courts charge private parties to help recover the costs of providing online trial platforms?

In civil cases, parties are already required to post jury fees, court reporting fees, and other costs that some courts assume for mounting trials. It would not be difficult for courts that choose to provide technology solutions to charge civil litigants a fee for their use.

In civil cases, private trial technicians, like some of the authors of this chapter, are also an important resource for parties and the courts. In civil cases where the parties have sufficient resources, the parties can and do provide the court, litigants, and jurors with technology solutions. We also serve as formal and informal consultants for judges and courts, sharing our best practices and teaching courtroom staff how to manage the court's technology resources. Courts and state governments might benefit by developing more formal public-private partnerships with trial tech companies that place the burdens of maintaining and updating technology on these private companies.

The Hidden and Not So Hidden Costs of In-Person Trials

We already spend a lot of money on physical courts and in-person trials. Entering 2021, California Governor Gavin Newsom signed a budget that includes $2.1 billion over five years to pay for new courthouse construction.[11] The costs of maintaining courthouses is significant as well, as are the personnel costs of running in-person trials.

Jim McMillan is a technology consultant who works with the National Center for State Courts (NCSC) and has been involved in bringing technology to courtrooms around the country for more than 20 years.

---

11. Maria Dinzeo, *California Courts See Funding Boost in Budget Proposal*, Courthouse News Service, Jan. 8, 2021, https://www.courthousenews.com /california-courts-see-funding-boost-in-budget-proposal/.

When he spoke with us, in response to the claim that online trials are too expensive, he did not wait a moment to reply:

> I reject that because in-person trials are very expensive. And I don't see that there's any significant cost displacement, there's no real big jump in between the two. You're displacing your physical costs like parking and you're replacing it with technology. I think the technology is probably cheaper than flying people and driving down the parking, dealing with all the security at the courthouse. Security issues! Do you know how much cost we have built in securing the physical courthouse facility? It is massive! And it is 24/7. You can never leave that thing open. It is just huge. If you want to do an honest comparison, find out what the courthouse security budget is and then say that, OK, this is what this is and this is what the video, what the web conferencing cost is, I think that the courthouse security is probably higher . . . you know, people cost money.[12]

David Slayton, administrative director of the Texas Office of Court Administration, echoes McMillan in comments he made to a reporter from the *Texas Observer*, noting, "If you look at it from the judge's perspective, there are a lot who have to travel. We have a judge in West Texas who travels four hours each day. Now, of course, they can log on and have that hearing without the four hours of travel."[13] Judges' time costs money; time spent on tasks like travel is also often wearying, as anyone who travels often for business knows.

In 2010, McMillan spearheaded a survey of videoconferencing use in state courts around the country.[14] More than 40 percent of state courts listed funding as an issue impacting their adoption of videoconferencing technology. However, that same survey reports that "time, staff, and fuel savings were cited as VC [videoconferencing] technology benefits," and while most courts had not quantified cost savings, those who had reported savings of "$31 million since inception ([Pennsylvania]), 30% of travel expenses ([Utah])," and other lesser savings from other states.

---

12. Interview by Josh Splansky and Marc King with Jim McMillan, Principal Court Technology Consultant, NCSC (Apr. 22, 2021) [hereinafter Interview with Jim McMillan].

13. Sundaram, *supra* note 10.

14. Jim McMillan, Videoconferencing Survey 2010 Results: National Center for State Courts, Presentation (Sept. 2010), https://www.ncsc.org/__data/assets/pdf_file/0013/15052/videoconferencing-survey-3.pdf.

These pre-COVID survey results reflect the complex truth that incorporating new technologies causes shifting and reallocation of existing resources, phasing out of some staff roles, and the creation of new ones, all of which unfold over time and a series of budget cycles.

In his interview, McMillan noted that when COVID-19 began, court budgets were not set up to account for online trials, and "it really takes 12 to 18 months to run through the court budget process. So right now, we're finally catching up on that. . . . It just takes time for the court budget process to go through."[15] He has seen courts address the budget issue by using civil trials where the parties can afford to bring in private trial technicians as an opportunity to experiment and learn best practices, to then prepare for criminal cases. Indeed, some of the authors of this chapter have been participants in that process in courts around the country.

Why Budget for Online Trials When the Pandemic Is Almost Over?

Some might ask, why bother to incorporate online trials into court budgets when, soon enough, the pandemic should be over, COVID-19 tamed, and we can all go back to normal? There are a number of reasons why we believe this would be misguided policy.

First, the pandemic has revealed the fragility of court systems that have no backup plan. The pandemic may or may not be over in the next budget cycle; regardless, it is clear that for the foreseeable future, we have a significant risk of facing similar situations locally, regionally, and nationally, whether due to pandemics, large-scale natural disasters, or human threats. Wildfires, floods, hurricanes, tornadoes, and epic snowstorms have all wreaked havoc with court services in various parts of the country over the past few years. Many people have forgotten that when 9/11 happened 20 years ago, most federal and state courts were shut down for many days. The pandemic has created an opportunity as well as pressure for courts around the country to create new practices that increase ease of access to the courts and ensure that the justice system can continue to function even when everyday life and in-person meetings are disrupted or impossible. We think the needs of justice demand that this process continues, not come to a halt because one pandemic ends.

Second, making the use of videoconferencing technology routine in trial holds the potential to cut costs and improve the delivery of justice in many ways. For example, as Jim McMillan said:

15. Interview with Jim McMillan, *supra* note 12.

> I personally think it's just crazy not to have the video web conferencing for normal work. So, for example, for years I've been saying that it's ridiculous to have the crime lab technician come and sit in a courtroom for days waiting for their expert testimony when they should be back at the lab cranking through the rape kits and all that sort of thing, trying to get this stuff done [when] everybody's complaining about the backlog in the crime lab.[16]

As we discuss extensively elsewhere in this volume, bringing trials online is not an all-or-nothing proposition. There are many ways to create hybrid trials that maintain in-person courtroom proceedings at the center but then allow for remote testimony of experts, as McMillan suggests; online hardship screening; online jury selection; and remote testimony of fact witnesses who are unable to attend in person.[17] We do not believe that a return to a society where COVID is under control should stop innovative uses of online technologies to save people's time, save staff costs, and improve the efficiency of justice.

Third, the evidence we present in this volume is that incorporating hybrid and full online trials as routine options has the potential to substantially increase both access to justice and the quality of justice of American courts—and the social and political costs of failing to incorporate technology into the administration of justice are likely to be much greater than the monetary costs of adapting existing technology to allow for online trials.

Saving money and efficiency are not the only concerns here; of course, it is important to ensure that online trials or uses of online technology in trials ensure fair trials and due process. Here, we are simply pointing out that those doubters who complain that online trials are too costly fail to see significant ways in which online trials can save time and money—and, may it be said, people's patience and spirit, which can often be worn down by the hurry-up-and-wait freneticism demanded by in-person trials.

Reimagining the Courtroom

Public-private partnerships between the courts and technology providers strike us as worthy of consideration partly because the biggest and most

16. *Id.*

17. In fact, traditionally subpoena power has been constructed based on geography because it has been considered unfair and improper for a court in one jurisdiction to compel someone living far away to appear at an in-person proceeding when they are not a party. If a witness can testify with relative ease from anywhere in the world, this could change our view of the power that courts should have to compel testimony by people who have relevant evidence.

expensive technical challenge that courtrooms face in moving to online trials is that existing courtrooms have terrible design for high-quality sound and video recording. They have been designed, like live theatres, for live performance and physical presence—not for high-quality studio recording and broadcasting.

If we decide as a society that online trials are worth offering, we need to design a new kind of courtroom, one that is more like a radio or television sound and recording studio, where the judge and attorneys, or judge, attorneys, and witnesses, gather to broadcast to jurors and the public; or where judges and attorneys gather to conduct an online jury selection with jurors logging in from home. This would be a good place for a public-private partnership, with courts providing space and private technology partners contracted to provide and maintain all needed equipment, software, and training. Managed well, this could keep courts on the cutting edge of technology while keeping costs down.

The image of traditional courtrooms is so strongly imprinted on our minds that creating "broadcast" courtrooms will no doubt seem to some a horror and aberration. It is true that the technical demands of online trials will drive a cascade of changes to trials and trial performance that are and will be disruptive. It is not evident that these changes will be bad.

We can see already some ways in which the shift to online trials creates benefits. For example, online presentation levels the playing field between men and women in certain ways, and between taller and shorter attorneys, as physical height and size do not have the same impact through a camera as they do in person. Neither jurors, litigants, nor witnesses can be intimidated or impressed through sheer size and physical presence—something we have seen attorneys and parties take advantage of in courtrooms many times. Attorneys with softer and quieter voices come across video at the same volume as attorneys with loud-booming voices.

Rob Rosenberg, one of the trial technicians we interviewed for this chapter, told us a story of one particular client,

> who said, "I want to thrash them while they're on the stand and I can't thrash them if I'm doing it over the Internet." I thought that was so foolish and felt like: That's literally the only edge you think you have in trial—whether or not that you can be aggressive and overly assertive towards a witness?[18]

As this story reflects, the shift to a new medium is disorienting to attorneys who have been successful in using their physical presence to lend

18. Interview with Rob Rosenberg, *supra* note 5.

to their performance; old dogs will have to learn new tricks. Young dogs will have new opportunities.

Costs of Data Storage

We hope that we have shown that the typical concerns voiced about the costs of online trials are off base. There is one cost, however, that is rarely raised by the public or the bar that is of concern to information technology (IT) professionals like Jim McMillan—and that is the huge cost of ongoing data storage of video files. As anyone who has taken videos on a phone quickly realizes, video files are huge and quickly gobble storage space. So much storage is now on the cloud, and cloud storage is provided by a small number of providers—Amazon is the leader here—who control the costs of data storage. Keeping video files of online trials indefinitely is not an option, as the data storage demands will quickly grow beyond courts' capacity to pay for that storage. This is not a new issue, as courts already have been holding some proceedings online and maintaining video archives, but the data storage challenges will grow exponentially and rapidly with the advent of online trials. Storing files locally off-line on hard drives may eliminate recurring costs, but would require personnel to back up and maintain the data. Courts and parties need to begin considering storage and retention policies now to ensure that budgets for data storage do not grow out of sight and out of control, as well as to address the privacy and security concerns that we discuss later in this chapter.

"Costing Too Much" Is a Value Judgment, Not a Fact

Battles over budgets are real, and government entities do have limited resources. Technology requires ongoing expenditures to support software licenses, technology purchases and upgrades, hiring of technical staff, and technical training. Court budgets will need to continue to evolve and adapt to these new realities. However, stating that online trials cost "too much" is a value judgment, not a statement of fact. Much of the software, hardware, and staff that are needed to support online trials are already in place. Additional monetary resources may be needed to support online proceedings as a routine option, including funding for training or retraining of court staff to provide the needed support to allow parties and the public to access trials. Money and resources follow commitment and common will. Narrowly focusing on funding issues misses what we think are the larger and more important questions:

- Can using online technologies help make justice more accessible and available to Americans?

- If so, what kinds of cases are best suited to online adjudication, or what portions of jury trials are best suited to online execution?
- What improvements do we need to off-the-shelf technologies to create a better trial experience?
- How do we need to reconceptualize the physical spaces of courtrooms to accommodate online proceedings?
- How do we need to retrain existing staff and what kinds of new staff roles are needed to support online trials?
- What are the costs to upgrade court systems to be able to provide high-quality online trial experiences, and how should we fund them?
- What kinds of public-private partnerships might be created to develop the technologies and policies needed to support online trials?

Party Technology Costs

In addition to having special constitutional requirements with regard to how the proceedings are handled, criminal trials most often involve defendants who have limited resources. They are also often in prison without access to computers, cell phones, or the internet. The state generally assumes most of the costs. The technical requirements to allow participation of a defendant in an online trial are relatively modest, however. Prisons could easily build a soundproof studio equipped with high-quality video and audio recording and streaming for a reasonable cost—a cost that could be funded from the savings from no longer needing to secure and transport prisoners physically to courthouses.

For this reason, we do not see cost as a prohibitive factor. The funding challenge is more a cultural and moral one, that taxpayers and politicians are often reluctant to provide resources to defendants in criminal cases. Even though we have a legal system that presumes the innocence of criminal defendants, most people believe if someone has been arrested and is in jail, they have done something wrong. While the state may be required to provide access for criminal defendants in online trials, there is good reason to be concerned about the quality of access that might be offered. States are required to provide legal defense to criminal defendants too, but this requirement is more often met in the breach.

Communication research[19] and commonsense experience show that the quality of video and audio recording and the quality of the internet

19. Several studies suggest that parties and witnesses appearing on-screen may receive less empathy from viewers than those testifying live. *See, e.g.*, Shari Seidman

connection matter in how people come across in video and over the internet. Criminal defendants who must appear in conditions of poor lighting, poor audio transmission, poor internet, and poor-quality backgrounds and surroundings will suffer in the judgments made of them. No jury instruction can cure the differences in subconscious impressions that will be created. If online criminal trials are allowed or required, as a society, we must provide all criminal defendants with the technical resources needed to present themselves properly—or we allow unequal justice.

Defendants also have the right to counsel, and one of the biggest concerns expressed by criminal defense attorneys about online trials during COVID conditions, where counsel and defendant would broadcast from different locations due to social distancing requirements, is that they would prevent a defendant from consulting with counsel.[20] Once the pandemic is past, however, there is nothing to bar counsel from being in the same room with their client, together with a sheriff's deputy or a bailiff or some other security officer at a distance or in an adjacent area.

Civil trials are a different animal. Civil trials involve private parties who usually have access to computers or smartphones and the internet.[21] Civil parties also may have the resources to provide equipment to the court and to jurors and to hire trial technicians who can assist with configuration, troubleshooting, technical training, and courtroom presentation.[22] As we have seen in the past year, bringing civil trials online has been much easier and it may well be for the next few years. For that

---

Diamond et al., *Efficiency and Cost: The Impact of Videoconferenced Hearings on Bail Decisions*, 100(3) J. Crim. L. & Criminology (1973–) 869 (2010); Ingrid V. Eagly, *Remote Adjudication in Immigration*, 109 Nw. U. L. Rev. 933 (2014); Penelope Gibbs, Transform Justice, Defendants on Video—Conveyor Belt Justice or a Revolution in Access? (2017); Sara Landström et al., *Witnesses Appearing Live versus on Video: Effects on Observers' Perception, Veracity Assessments, and Memory*, 19(7) Applied Cognitive Psych. 913 (2005); Holly K. Orcutt et al., *Detecting Deception in Children's Testimony: Factfinders' Abilities to Reach the Truth in Open Court and Closed-Circuit Trials*, 25(4) Law & Hum. Behav. 339 (2001).

20. See, for example, National Association of Criminal Defense Lawyers, Criminal Court Reopening and Public Health in the COVID-19 Era: NACDL Statement of Principles and Report 5 (2020), which extensively discusses the ways in which virtual proceedings may be impacting the Sixth Amendment rights of criminal defendants.

21. See Chapter 8 in this volume for data on the availability of computing technology and broadband internet access in the United States.

22. Very few cases go to trial anymore in either our civil or criminal courts; in the civil courts, the cases most likely to go to trial involve disputes involving large companies and individuals with substantial financial resources.

reason, the majority of full online trials have been civil cases. It may be that some courts decide that only civil trials are suitable for online trials.

Many civil litigants have insurance that helps pay some or all of their costs; insurance companies should be encouraged to include coverage for the technology and technology assistance needed for online trials. Given the potential for online proceedings to save substantial time and money, we think insurance companies will naturally begin moving in this direction.

Problems are more likely to arise from situations where, as online proceedings become routine, insurance companies refuse to pay for in-person trials in situations that their attorneys believe merit them. It could create situations like those we see in medicine, where a doctor believes one treatment is best for the patient but the insurance company denies coverage and presses for another, cheaper treatment option. This is a policy issue, not a technical issue, that may need to be addressed by legislatures and policy makers.

Family law hearings and trials are more likely to involve litigants who have limited resources and may find it challenging to participate in online proceedings. According to the survey conducted by Judge Pamela Gates and the Maricopa County court system in Arizona, the number one problem jurors cited as interfering with online jury service was not lack of access to technology, but lack of access to a quiet space where the juror could be without interruption for the length of a court proceeding.[23] Some did lack computers or high-speed internet access as well. During the pandemic, courts holding online trials have dealt with these challenges by creating workstations using existing court computer resources where jurors can go and access the proceedings. It will surely take some creative interior designers, but it seems existing government and court office spaces can be reconfigured to provide this access on a larger scale if necessary. Public libraries already offer computer and internet access; it would not be difficult to create dedicated rooms in libraries set up for participation in trial, which would create a geographically wide range of access.

Juror Technology Costs

Holding full online jury trials requires not only that courts be prepared to host and broadcast trials online; it also requires that jurors have

23. See the discussion of data from Maricopa County, Arizona, in Chapter 8 of this book.

sufficiently robust internet and computer technology to participate in the trial and know how to use it. This raises two questions:

- Do we want jurors using their own personal devices to participate in trials? Is this sufficiently secure and easy to manage?
- What do we do about jurors who lack the needed technology?

Managing the costs of providing technology and training to jurors is more challenging both short-term and long-term than managing the costs of court technology—particularly for criminal courts, where the cost challenges are different than in civil courts. Part of the challenge is logistical: courts are stable central nodes in the trial process whereas jurors are a constantly shifting network of individuals, geographically dispersed and without a particular commitment to the court function or location. There are solutions, however. Those solutions are discussed in detail in Chapter 8 of this volume, on representativeness, so we will not repeat them here, but will simply note that trial courts and parties have found solutions that work in trial courts across the country.

"The Learning Curve Is Too High"

With full online trials or online jury selection, there are two aspects to the learning curve:

- The learning curve for courts, court staff, and counsel who are professionals engaged with trial on a regular basis
- The learning curve for litigants, jurors, and witnesses, who are occasional or one-time participants in the trial process

The evidence of the past year is enough to disprove the naysayers who suggest that the learning curve is too great for courts and court staff. Courts across the country, including those in some of our most populous states (Florida, Texas, California), have come together to provide online trials for their citizens. Judges around the country have stepped up to teach themselves and then turn around and teach their colleagues, including Judge Emily Miskel in Texas, who just won the William H. Rehnquist award for her work. Court administrators like David Slayton of Texas have also stepped up to provide the training and support necessary to train courtroom staff as well as judges.

As Chapter 8 discusses in detail, the data shows that a large majority of Americans have access to both computing devices and broadband internet service. Some, however, do not, and those people tend to be among the most marginal and socially disenfranchised groups—including

Native Americans on reservations, resident aliens and immigrants, older people, Black Americans, and Hispanic Americans. Providing access to technology is usually not enough for people who lack access due to poverty and social disenfranchisement; because they do not routinely use this technology in everyday life, the parties and courts must also ensure that they are given technical support and training so that they know how to use the technology.

Courts have already been providing training with parties who are called to online hearings and arraignments through existing technical staff or by training courtroom staff to give instruction. In this past year, the judges we interviewed all spoke about themselves stepping in to help jurors with technical issues at times as needed. In civil cases that hire trial technicians to help mount an online trial, the technicians can take charge of "tech checks" and training sessions with jurors to ensure jurors know how to use the hardware and software properly.

In an ongoing way, the people best situated to conduct training are a "technical bailiff," trained and employed by the court; or a private trial technician retained by the court or the parties to provide online trial support services. The best practice is to have group tech checks and training sessions for online jury selection, with individual follow-ups as needed, and one-on-one tech checks and training sessions with jurors once they are chosen to sit on a panel. One-on-one training sessions with parties and witnesses should also be arranged. The evidence that we have from our interviews and our own collective experience mounting more than 100 online trials is that while best practice is to hold tech checks with all online trial participants, a relatively small and manageable number of jurors require extensive training and technical support.

"Managing Online Jury Selection Is Too Hard"

Our first response to people who say managing online juries is too hard is simple—have you ever picked a jury in state court?! Managing *in-person* juries is too hard!

The average size of a jury pool is 40 jurors. In longer and more complex cases, or cases with high levels of publicity, jury pools may reach sizes of 120 jurors or more. Managing all of these bodies, ensuring that jurors are seated in order of their juror number so counsel and the court can keep track of who's who, is no simple task and it typically takes at least an hour initially when selection begins, and at least 20 to 30 minutes every morning on subsequent days of voir dire. Managing juries online, from management and technical points of view, is actually much easier once courtroom staff are familiar with the software tools.

The judges we have interviewed have landed on a common set of best practices for managing online jury selection:

- Have jurors complete a screening questionnaire and a supplemental questionnaire online.
- Screen out jurors who clearly qualify for hardship or cause challenges based on those submissions.
- Schedule follow-up individual hardship questioning with specific jurors as necessary.
- Once the pool has been whittled down, schedule jurors to appear for voir dire in panels of between ten and 15 jurors for two-hour blocks. Fifteen is the limit that can appear on-screen for quality viewing.
- Schedule follow-up individual questioning as may be necessary.[24]

Handling jury selection this way, unlike in in-person jury selection, the entire jury pool is not present for the entire voir dire process. This is a difference, but is it a problem?

It is not unusual in criminal cases for all questioning to be handled in camera, one by one, in which case jurors do not hear questions or answers for any other jurors.

We are not aware of any studies that examine the impact on jurors who are not initially "in the box" for questioning of sitting through the voir dire process. Our experiences in court suggest that there are some benefits from jurors sitting in the gallery, listening to the voir dire process, and learning about what qualifies as bias. Typically, voir dire does begin to move more quickly as jurors learn the process and understand the case issues better. Our experience is also that judges often truncate questioning by attorneys in later rounds of voir dire, based on the assumption that jurors in the gallery have heard and understood the questions and do not require the same level of questioning as jurors earlier in the process. As a result, we have seen that it can be harder to flush out jurors who are "deeper" in the pool who have biases and are holding back or are reluctant to speak. We are not aware of any reason why organizing jurors as we describe in this online method might impede the selection of a fair and impartial jury.

---

24. Interview by Marc King & Josh Splansky with Judge Sean O'Donnell, King County Superior Court, Washington (Apr. 21, 2021) [hereinafter Interview with Judge O'Donnell]; Interview by Marc King with Judge Emily Miskel, 470th District Court, Texas (Apr. 9, 2021); Pamela S. Gates et al., *Virtual Juries: We Can but Should We? And if So, How?*, 47 Litigation 1 (2021); *see also* Chapter 15 on jury selection.

"You Cannot Pick a Jury through a Screen"

The second frequent objection about online jury selection is that technology is no substitute for presence, and too much is lost without the ability to observe body language and feel a juror's presence and reaction to counsel.

It is unquestionably true that human beings react differently to each other when we are physically present to each other versus viewing each other on a screen. When in physical presence, we have a host of subconscious physiological responses beyond our control that impact our reactions to others. Scientific and social scientific research over many years and across disciplines suggests that templates for our physiological responses are laid down in early life and we are conditioned to respond more favorably to those who are socially, culturally, and genetically more similar to us. A reasonable conclusion is that physical presence is therefore a mixed bag as far as delivering justice and picking fair juries go.

Judge O'Donnell in fact opined that he *prefers* remote jury selection because it makes it harder for judges and counsel to make decisions based on biased appeals to body language and gestures. In Washington state, where he practices, the court adopted rules in 2018 that point out:

> The following reasons for peremptory challenges . . . have historically been associated with improper discrimination in jury selection in Washington State: allegations that the prospective juror was sleeping, inattentive, or staring or failing to make eye contact; exhibited a problematic attitude, body language, or demeanor; or provided unintelligent or confused answers.[25]

The rule continues:

> If any party intends to offer one of these reasons or a similar reason as the justification for a peremptory challenge, that party must provide reasonable notice to the court and the other parties so the behavior can be verified and addressed in a timely manner. A lack of corroboration by the judge or opposing counsel verifying the behavior shall invalidate the given reason for the peremptory challenge.[26]

Judge O'Donnell reported that his online jury selection this past year resulted in more diverse jury pools and more intimacy with the jurors,

25. WASH. STATE CT. GEN. R. 37(h)(i) (2018), https://www.courts.wa.gov/court_rules/?fa=court_rules.display&group=ga&ruleid=gagr37.
26. *Id.* 37(i).

and that the quality of juries overall was similar to what he has experienced with in-person juries. From his vantage point on the bench, technology has not stood in the way.

For years, judges and attorneys have relied on their instincts in jury selection, much like police and prosecutors have relied on eyewitness testimony to catch and prosecute people for crimes. It is not clear that these gut instincts about other people are reliable, as much as we might believe them. In fact, there is a considerable body of evidence that judges and police officers believe they are superior at judging credibility but, at best, are unable to detect lying any better than lay people—which is not very well—and, at worst, make judgments marred by confirmation bias.[27] We tend to be drawn to people who we feel are "like us," but that is no guarantee of honesty or lack of bias. Removing this element from jury selection by viewing jurors through a screen may actually make the selection process more objective and fair—although it does prevent counsel from determining if a juror may be physically "turned off" by their presence, which can be a factor if jury selection is online but the trial is held in person.

As far as observing jurors' body language and gestures, online jury selection puts every juror's face front and center on the screen, at the same distance, and allows for more even viewing than a jury box in the courtroom, where the tall juror in the front row may block your view of the short juror in the back—although, again, research suggests that body language may mislead people in our credibility judgments as much as it informs.[28]

When the OCP ran the first demonstration trial in June 2020, we interviewed and surveyed our mock jurors and the participating attorneys about their experiences. Judges O'Donnell and Miskel have also conducted interviews with both jurors and attorneys who have participated in their online trials. We all heard the same types of comments: attorneys often expressed surprise at how well the process worked and, because they saw into jurors' living spaces, felt that they had better insight into the jurors' everyday worlds. Jurors talked about the convenience and ease

27. *See, e.g.*, C.A. Meissner & S.M. Kassin, *"He's Guilty!": Investigator Bias in Judgments of Truth and Deception*, 26 LAW & HUM. BEHAV. 469 (2002), https://doi.org/10.1023/A:1020278620751; S.M. Kassin et al., *"I'd Know a False Confession if I Saw One": A Comparative Study of College Students and Police Investigators*, 29 LAW & HUM. BEHAV. 211 (2005), https://doi.org/10.1007/s10979-005-2416-9; Jill Yarbrough, *The Science of Deception Detection: A Literature and Policy Review on Police Ability to Detect Lies*, 3 J. CRIM. JUST. & L. (2020).

28. *Id.*

of the process, and also said that they felt more comfortable with some of the intimate questions speaking from home rather than in a public courtroom full of strangers. We and the judges all observed that jurors were more relaxed and conversational in their answers. Picking a jury online is different but it offers some significant advantages and it can be done.

"Monitoring Online Juries Is Too Hard"

Well, naysayers respond, maybe you can pick a jury online, but managing and monitoring a jury during an online trial is too hard.

In a Law360 article, two attorneys from a leading law firm express some common sentiments about the challenges of managing juries in online trials:

> In a normal courtroom setting, courtroom staff and judges work together to ensure that jurors are present, safe, and attentive. Juror interactions and access to the "outside world" are also more easily managed when the judge and jury are in the same location. These controls prevent, or at least impede, jurors from being influenced by externalities. There would be no way for a judge to subtly wake a juror before he nods off during testimony, or to know whether there were individuals (or electronic devices) off-camera influencing jury members during trial or during the deliberation process.[29]

The setup here is typical of opinions offered about online trials—the authors assume that the pre-COVID norm worked to manage juries and that there is "no way" to carry out management and oversight of jurors in an online setting.

We do not mean to single out any particular firms or authors. The sentiments expressed above are widespread and typical, and they are based to some extent on commendable concerns and worries about the integrity of the trial process.

However, sober analysis of pre-COVID-19 experiences in court together with data about the rapid evolution of jury management practices in online trials this past year suggest that both assumptions are off the mark.

29. Paula Hinton & Tom Melsheimer, *The Remote Jury Trial Is a Bad Idea*, LAW360, June 9, 2020, http://www.law360.com/articles/1279805.

Pre-COVID, in-person jury trials have faced significant challenges managing juror behavior. During COVID, courts that have held online trials have been developing effective ways to ensure juror attention and monitor juror conduct—although to be sure, practices are evolving and the technology would benefit from evolving to account for the particular needs of trial courts.

We think such sentiments also ignore the most important factor that ensures the success and integrity of jury trials—social trust. Ultimately, communication technologies are only extensions of human relationships. New technologies bring new challenges; the solutions to those challenges are sometimes technical but are more often social and cultural. The deeper issue is how to create a sense of commitment and trust between jurors and the court that results in understanding of and adherence to court rules.

To that point, as with the author above, many observers give more credit to the quality of juror oversight in in-person trials than experience suggests they merit, assuming that mere physical presence is enough. In-person trials are no guarantee that jurors will be engaged and attentive during trial. Every trial professional has a story about a sleeping or snoring juror. Judges are trained to monitor jurors from the bench and will call for breaks if they see a juror's attention flagging, or may call out a specific juror who seems to be nodding or dozing off, but even monitoring from the bench cannot guarantee that a juror is paying attention and focused.

In addition, in-person trials do not guarantee that jurors will make their decisions based only on what happens inside the courtroom. During breaks in in-person trials, jurors can talk to fellow jurors about the case and observe and overhear counsel, witnesses, and parties as they engage and go about their business outside of the courtroom. Most of us working on this chapter have experienced in our careers cases that have been decided based on a juror observing the injured plaintiff leaving a bathroom stall without an indication of injury, or overhearing and misinterpreting a conversation between a defendant and counsel, or drawing a conclusion about an expert based on the way he acted in line at the courtroom cafeteria—which is why it is standard to tell both expert and fact witnesses during prep to act as if they are "on stage" from the moment they leave their house to go to court until the moment they walk through the door to come home at night. These out-of-court observed moments, however, are extremely unlikely to happen in an online trial.

One of our authors conducted post-trial interviews on a major jury trial in southern California where the jury deliberated for months, played

Scrabble during deliberation sessions, discussed newspaper articles about the case that some had read in violation of court orders, and talked about the impact that they wanted their jury decision to have on their community. Being in person mitigated against doing their jobs as jurors!

With regard to "off-camera" activity, when jurors in in-person trials are outside of the courtroom on breaks, in most courts,[30] they have access to their phones and devices and can readily search the internet for information about the parties and the case, regardless of the court's instructions. They can also blog, tweet, and post about their activities. The evidence is that both happen with regularity. In 2011, one of our authors worked on a major case in San Francisco in which two jurors were dismissed during jury selection because they violated the court's instructions and posted to social media about their participation in the trial.[31] In 2010, right around the same time, Reuters reported that "jurors' forays on the Internet have resulted in dozens of mistrials, appeals and overturned verdicts in the last two year"[32] and reported on instances of mistrials around the country resulting from improper juror research and communication during trial.

The conclusion any reasonable observer must make is that the challenges of managing jury service under conditions of new technology are not specific to online trials, even though the specific management challenges may differ between in-person and online trials.

So again, as with many issues we are discussing in this chapter, the baseline conditions are far from perfect. People presume that physical presence equals connection, engagement, and adherence to the rules, but that is not the case. Attention is never guaranteed. Judges and advocates must earn attention and respect. We could do better engaging jurors in in-person trials than currently happens.

That being said, participating in any kind of meeting or group event through videoconferencing is different from participating in person, and judges and attorneys must account for that. Even viewing witness

30. Many federal courthouses require jurors, non-attorneys, and the public to leave their electronic devices outside or with a marshal, but some allow them in. Most state courts allow jurors to bring their electronic devices with them into the courthouse. Jurors are of course required to turn them off during court sessions but they can turn them back on during breaks, and most do.

31. This happened in 2011, when Sarah Murray served as an in-court jury consultant during jury selection in *Rambus, Inc. v. Micron Technology, Inc.*

32. Brian Grow, *As Jurors Go Online, U.S. Trials Go off Track*, Reuters, Dec. 8, 2010, https://www.reuters.com/article/us-internet-jurors/as-jurors-go-online-u-s-trials-go-off-track-idUSTRE6B74Z820101208.

testimony via video is different from viewing testimony live and court proceedings are better if this is acknowledged and addressed.

Online jury selection and online trial participation place different demands on jurors, however. Online jury selection is much more interactive and it is difficult for jurors to remove themselves from participation. In fact, it is easier for a juror to hide in the jury box and remain quiet during in-person jury selection than it is for a juror to hide out during online jury selection, since every juror's face is equally present and visible to the judge and attorneys. The most serious challenge in online jury selection is technical—ensuring that each juror has access to a device and internet connection of sufficient quality to allow them to "show up" and be seen and heard as clearly as other jurors, and these challenges are manageable.

Monitoring jurors' participation in an online trial proceeding is more difficult. There are currently not good technology solutions for monitoring of juror attention. However, it is not clear that technical solutions are the best way to go. Employers currently use technologies like keystroke monitoring and programs that track browser and email traffic to monitor their employees' computer use, to ensure that they do not engage in unauthorized computer use during work hours. Theoretically, these technologies could be adapted for jury service. This kind of "big brother," panopticon monitoring both indicates and potentially breeds a lack of trust and respect in the work environment, however. Psychologists recognize that the drive for autonomy is a primary drive in human beings. We all want to have control of our own time and lives—jurors too. It is contradictory to have a system based on honoring and treating jurors and juries with respect but yet imposes electronic monitoring on jurors based on the presumption that jurors will be inclined to violate the terms of their service.

There are ways that judges and attorneys can create good conditions for juror participation. These include:

- Stressing how important it is that jurors pay full attention by reminding them of the ways that their decisions will impact the parties
- Letting jurors know that justice depends on their attention to the evidence and the law
- Asking jurors to mute themselves but keep their video on during breaks so that they continue to be "present" and connected
- Taking regular breaks, at least five minutes every hour, to allow jurors time away from the screen to refresh their attention in their local environment

- Periodic reminders at the end of breaks to jurors to close out of any other open windows and to place their full attention on the proceedings
- Recognizing that online proceedings are more tiring for our brains and eyes and limiting trial presentations to three hours per session

There is a small subset of jurors who will be bad actors, and for those, we may need to develop new laws that fine and punish people called to serve as online jurors who violate their oath and terms of service. It is best to design the system for the typical juror, however, not for the "bad apple" juror.

We recommend that any court holding an online trial have someone serve as "technical bailiff"—not only to orient and train participants in how to use the technology beforehand, but to manage the online trial process. This person's job should include helping the judge monitor juror attention and contacting jurors if they appear to be sleeping, disengaged, communicating with others off-screen, using other programs during trial, or engaging in any other improper conduct. Jurors can be contacted through one-on-one chat connections in programs like Zoom or Teams (which creates a savable record) and, if necessary, direct phone text or telephone contact.

These best practices point to the fact that the most effective solutions to technology challenges are rarely more technology—they most often are ways to activate core human social responses like conformity, trust, and social commitment under new conditions of interaction. We, as human beings, are social creatures. Most of us want to win the approval of others and show ourselves to be good people according to the terms of our society. In-person jurors sitting in a jury box can easily go out in the hallway during a break, ignore the judge's injunction, and search the internet on their phone for information about the trial—and the evidence is that some do. What stops a juror from doing that is not a software program or a setting on their phone. It is not a judge or bailiff looking over their shoulder. It is instead a sense of trust in the process and a commitment to upholding the laws and ideals of our polity. If jurors understand why they are being asked to conduct themselves in certain ways, if jurors are treated with respect and given to know the importance of their role in the process of justice, this goes a long way to encouraging proper juror conduct, whether in an in-person or an online trial.

"Document Handling Is Too Hard"

Documents are the lifeblood of trials, particularly civil trials. Managing documents has become increasingly complex, particularly in civil trials, since the inception of desktop computers, which have made creating and saving documents all too easy. Well before the advent of COVID-19 and online trials, the volume of documents involved in litigation has threatened to overwhelm litigation and the courts. E-discovery management companies have emerged to help manage the flood of documents; years of the lives of young associates have been lost to document review and management; and trial presentation specialists like some of our authors emerged to help digitize, organize, and present documents in trial. This flood of documents is part of what has pushed civil lawyers in particular towards graphic presentations in trial, as this high volume of documents is best handled by digitizing documents and creating databases to tag, sort, and manage them.

This is why, over the past 40 years, courts have moved more and more to digital and online document management and why digital documents are the standard in trial practice in both state and federal courts.

Presenting digital documents in an online trial presentation is no more difficult than displaying them on-screen in a physical courtroom. Through a program like Zoom, documents can be shown through a "screen share" function. Legal solutions companies like Veritext offer document management and presentation services that allow for side-by-side viewing of documents and live or recorded video.

Many jurors find it easier to see the details of a document on a computer screen eight inches away from their eyes than an image projected onto a screen 20 to 30 feet away from them, as is typical in many courtrooms. Some in-person courtrooms are equipped with personal monitors in front of each juror's seat; seeing evidence on a screen in the jury box is little different than seeing it on one's own laptop or computer monitor at home.

During deliberations, jurors need to have access to documents directly so they can review the evidence. Currently, there are three good methods that can be used for distribution of evidence to jurors:

- Documents can be uploaded on secure internet services and set to be viewed only, so jurors cannot print or download. Links can be sent to jurors via email or chat, and they can view on their local devices. The documents can be locked to prevent downloading and jurors can be given unique links to allow them access, so that access can be tracked and monitored.
- Evidence can be preloaded or pushed onto local devices distributed to jurors to use for the duration of the trial such as iPads or Chromebooks. The program or device can be set to disable

printing. Post-trial, these devices can be collected and wiped to ensure security and privacy.

- Jurors can be sent a hard copy set of exhibits to their home addresses that they are required to (1) deliver to a local shredding location to then provide proof of delivery to the court or (2) return to the court or to an agreed-upon location or party for shredding post-trial.

The biggest challenge that we have seen with online trial document management is that production of evidence cannot be left to the last minute. Getting exhibits properly organized and prepared for presentation and juror access takes planning, coordination, and time. In an in-person trial, you can walk into court and slap a hard copy document down on an ELMO document camera and get it before the jury; in an online trial, you cannot do that.

Handling documents in cross-examination, particularly for impeachment, also requires more planning and forethought and we have heard trial attorneys complain about this. This issue is addressed in other chapters in this volume, including Chapter 16 on testimony and Chapter 17 on exhibits.

One effective method of handling potential impeachment exhibits is to place each document in a sealed envelope, and then send the entire set of various sealed envelopes to the witness. The package should include a cover letter stating, "This package contains Exhibits XX through XX, each in an individually sealed envelope. Do not open until instructed." During testimony, the witness may be instructed by the court or counsel to hold the package up in front of the camera so all can see that it is still sealed. Then they can be instructed to open the envelope, holding it up in front of the camera in order to verify it has not been opened or tampered with previously.

Another and perhaps better method would be to simply screen share the impeachment exhibit, allowing the witness ample time to review. At this time, electronic copies may be sent to the court, opposing counsel, and the witness. Handling impeachment evidence in this manner minimizes the risk of unintentional (or intentional) unauthorized access to confidential material.

"Public Access Is Too Hard"

In America, a fundamental part of our jury system is that trials are open to the public and, except in limited circumstances, anyone can, as we would proverbially say, walk in off the street, enter a courtroom, sit in the gallery, and watch the proceedings. Public oversight is an important mechanism for keeping judges and juries honest. Prior to COVID-19, maintaining public access was easy—anyone who had identification and could pass through security could walk into a courtroom and listen to the trial proceedings.

The move to both socially distanced in-person trials and online trials has created challenges around ensuring public access.

Technically, these challenges are relatively simple to meet:

- Broadcast proceedings locally in the courthouse to viewing rooms where the public can gather in ways that meet public health regulations. These broadcasts can be via closed-circuit television set up in the courtroom or via internet services.
- Stream proceedings via services like YouTube, controlling the settings behind the scenes to prevent viewers from downloading, sharing, modifying, or commenting on the trial proceedings.
- Set up audio feeds that allow people to connect to the trial through audio only.

Governments around the country have been using closed-circuit television and dedicated cable channels to broadcast legislative and executive branch meetings for years; C-SPAN began in 1979. Technically, ensuring public access is easy and in fact broadcasting makes trial proceedings more easily available to more people; this is why the greater questions and challenges arise around security and privacy, issues that we address in the next section.

"Online Trials Are Too Insecure"

Early on, "zoom bombing" and unwanted intrusions into online trials were common and widely reported in the news. This created an impression that online proceedings were insecure and was seized on by critics as "proof" that online trials could not work. These "break-ins" were in great part the result of new users not yet understanding the security needs and the security tools of online videoconferencing. They were also partly the result of videoconferencing programs being used in new ways, for uses and audiences that they were not originally designed. Videoconference software companies quickly recognized the need to upgrade their security settings to account for new needs and made adjustments. Judges and trial technicians quickly learned how to optimize security settings of videoconferencing programs to prevent this kind of intrusion, and they stopped being a problem.

Judge Sean O'Donnell of the King County Superior Court was one of those involved in this early learning; he shared with us his thoughts and experiences with ensuring security of his online proceedings:

> I think platforms like Zoom are getting better, but that's in combination with judges who need to invest time taking steps and precautions to ensure the hearing or the proceeding is secure.
>
> I speak with some authority on this as a victim of a Zoom bombing during a hearing of mine. It was part technology and part my shortcoming as the operator of the hearing.
>
> Some of these Zoom bombs have been quite awful and I would never want to minimize the potency of one of those disruptions because they can be quite serious, terrible, long lasting in terms of just memories and the like. The worst-case scenario for something like that is you have a mistrial. We guard against mistrials in real court settings too, and this is just a different venue in which you need to guard or protect against that eventuality.
>
> There are simple things to do, like making sure you know who you're admitting into the meeting, ensuring that you are familiar with the security protocols, all of which are pretty easy to do.
>
> If you want your hearing to be accessible to the general public, you set up a YouTube page or an equivalent. I've done that.
>
> After my experience in the early days, I significantly limit who comes in on the Zoom call. I know everyone who's logged in. I lock the meeting, I lock the participants from being able to communicate amongst themselves. And I think if you were to ask the hosts of these video conferences like Webex or Zoom or Google, they would tell you that their security protocols are improving.
>
> To be sure, you still have hackers out there. People who want to disrupt and protest and the like, but to borrow an acronym, IRL—in real life, that happens too.[33]

Our own experience matches that of Judge O'Donnell. On a technical level, ensuring security of online trial proceedings is today a relatively easy task for anyone with even basic training in the use of online

33. Interview with Judge O'Donnell, *supra* note 24.

videoconferencing programs. In many ways, maintaining videoconferencing security is easier than maintaining the physical security of courthouses.

However, on a broader level, security of online trials is a thornier issue that touches on issues more general to our society's widespread adoption of visual recording and streaming technologies.

There are two aspects of global-level security management that bear attention and discussion to minimize hacking and disruption of proceedings as well as to address privacy needs of parties, witnesses, and jurors:

- Creating programs and cloud storage services that at the programing and management levels ensure the security and integrity of the program and the digital information that is being recorded and streamed
- Considering best practices for recording and broadcasting trial proceedings that allow courts to prevent public release of protected information, and to ensure the privacy of jurors, parties, and witnesses

Program and Data Security

American governments have been using computers and computing technology for years now; many of the programs that define our IT landscape now, like the internet, were inventions of the federal government. With regards to global security of programs and data, both the federal and state governments already publish security standards that vendors are required to meet.

Federal courts are required to use only programs and products that meet the Federal Risk and Authorization Management Program (FedRAMP) standards, which are standards set for secure cloud services. To meet this standard, programs and products must go through a multistep process of review, testing, and authorization. The standards are stringent; currently, according to the fedramp.gov website, there are 34 programs ready to be used, with 54 in process and 220 authorized.

States each have their own set of standards and they vary in the stringency of those standards and the quality of their technical review and oversight of programs. As is true with other technologies, such as voting machines, this variation in state standards and oversight can create complexities in ensuring the security of governmental processes and information. However, in practice, what tends to happen is that states with more robust standards and oversight, such as California, Texas, and Washington, set the standard for commercial companies like Zoom and Microsoft that then create government versions of their videoconference programs to comply and other states then follow in their wake—much as happens with textbooks for public schools, voting machines, automobiles, and other technologies and products that draw government oversight.

Zoom, Microsoft Teams, Webex, and other commercial videoconferencing providers all have versions that are created specifically for government use that adhere to the Health Insurance Portability and Accountability Act (HIPAA) and other federal and state government security protocols. Any judge, attorney, or party who is considering participating in an online proceeding or trial should ensure that the court or trial technician who is providing the technical platform is running the government-approved version.

These security features do not guarantee that hacking, manipulation, and unauthorized access to data will not occur, as is true with any activity carried out using the internet. This is unlikely to be an issue with routine trials, but could well be an issue with high-profile trials or trials that involve valuable technology or information. For this reason, courts, the bar, and parties should carefully consider what kinds of trials and cases might be best suited for online methods. The fact that online trials can be done, and done well, does not mean that every case is well suited for online trial. There is a trade-off between convenience and access on the one hand, and security and privacy on the other. It would be a good task for an American Bar Association subcommittee of the Trial Practice Division to bring together judges, attorneys, trial technicians, court administrators, and other stakeholders to create an algorithm for deciding if a case is suited for online trial, as well to create best practices for how to record and stream online proceedings to ensure appropriate privacy and security of participants, in both the short and long term.

Reconciling Public Access with Privacy and Security Concerns

A core principle of American justice is that court proceedings, in most instances, are open to the public. The belief is that public oversight helps keep justice honest and public access helps educate citizens about democracy and justice work. In practice, in-person trials that are not allowed to be broadcast limit the public viewing of trials because people must show up to the courthouse and find a seat in a gallery, which can accommodate a limited number of bodies. With high-profile trials like the recent Derek Chauvin trial, there are never enough seats for observers—even without social distancing restrictions in place. That is why Judge Cahill, the judge in the Chauvin case, decided to grant access to cameras to broadcast the entire trial live—something that had never before been allowed in a Minnesota court.[34]

If you do not show up live for an in-person court proceeding and it is not broadcast or recorded, it is difficult to get much information about it. Media reports about even the most high-profile trials are thin with information and evidence. While technically trial transcripts are public record, they are not easy to access or distribute. For even a litigant to obtain access to a trial transcript, generally one must approach the court or the court reporter, make a formal request, and pay a fee. For most types of cases, court reporters are not required to keep transcripts or recordings indefinitely; courts also vary in their policies for how long they will keep both paper and digital transcripts. This means that parties, witnesses, and jurors have had, practically speaking, significant privacy because of the barriers to obtaining detailed records of proceedings.

By contrast, video recordings of trials that are streamed on platforms like YouTube, which are commercially owned broadcast media, are readily available for anyone with access to a computer and internet access to view. Although you can create settings on YouTube and other platforms that prevent recording in that platform, there are many third-party programs available that allow you to take video and still "captures" of a computer or smartphone screen, which means that a recording of a trial could be available indefinitely on the internet for access and use by anyone, without courts being able to exercise any control over how those recordings are used. Advances in video technology, facial recognition technology, and the advent of "deep fakes" means that video proceedings have the potential to be used in nefarious ways.

This is where we think the most serious and reasonable security and privacy concerns exist with online trials. Young adults who grew up with social media learned the hard way that photos you post of yourself partying and in compromising positions do not go away on the internet. The potential for video footage of trials to live forever and come back to haunt participants in unexpected ways, or to be used in political attacks and social media political misinformation campaigns, is real.

The issue of cameras in the courtroom is not new. Television and cable news outlets have sought to bring cameras into the courts since the 1970s. Most courts have refused or severely restricted the use of cameras

---

34. Jack Dutton, *Who Is Judge Peter Cahill? George Floyd-Derek Chauvin Trial in Focus*, Newsweek, Mar. 29, 2021, https://www.newsweek.com/who-judge-peter-cahill-george-floyd-derek-chauvin-trial-1579515.

in courts because of privacy and security concerns for jurors, parties, and witnesses. The federal court system has run two pilot programs testing the use of cameras in the courtroom; both resulted in recommendations to severely limit cameras in the courtroom during jury and bench trials, although the federal courts now allow cameras in court for appellate sessions. The current policy for the trial court level is as follows:

> A judge may authorize broadcasting, televising, recording, or taking photographs in the courtroom and in adjacent areas during investitive, naturalization, or other ceremonial proceedings. A judge may authorize such activities in the courtroom or adjacent areas during other proceedings, or recesses between such other proceedings, only:
>
> 1) for the presentation of evidence;
> 2) for the perpetuation of the record of the proceedings;
> 3) for security purposes;
> 4) for other purposes of judicial administration;
> 5) for the photographing, recording, or broadcasting of appellate arguments; or
> 6) in accordance with pilot programs approved by the Judicial Conference.
>
> When broadcasting, televising, recording, or photographing in the courtroom or adjacent areas is permitted, a judge should ensure that it is done in a manner that will:
>
> 1) be consistent with the rights of the parties,
> 2) not unduly distract participants in the proceeding, and
> 3) not otherwise interfere with the administration of justice.[35]

What is new with online trials is the use of the internet to stream trials; the use of cloud data facilities to store video records; and the potential of trial records to be obtained by people socially and physically remote from the trial to use for nefarious purposes.

Alternative ways of ensuring public access have included courts providing audio access only. If someone really wants to attend the court sessions, it is possible, but only to listen. This is usually done by providing a call-in number and they can dial in to the court call or the court's own internal conference call system.

35. U.S. Courts, *History of Cameras in Courts*, https://www.uscourts.gov/about-federal-courts/judicial-administration/cameras-courts/history-cameras-courts (last visited Nov. 6, 2021).

Providing audio-only streaming or broadcast services may be the best way to provide public access while also still protecting participant privacy and security. The issues of protecting privacy and ensuring security are serious enough, though, that this merits further work and dialogue among stakeholders.

The U.S. District Court of Massachusetts has implemented a method of requiring members of the gallery to register with the host in advance of a trial day to gain access to the broadcast, thus curtailing anonymous viewers and potentially warning of consequences for those who do not comply with the court's rules. As courthouse IT Project Manager and Administrator Phil Doreau and Audio/Video (A/V) Specialist Lynne Shannon, explain:

> We created a registration site. So, everybody is supposed to provide their name and info before they get a link to see the proceedings so that we have some kind of database to tie back if we have any issues with any individuals. And we don't allow anyone into the hearing or trial if they haven't registered. We have somebody that checks them in before they're allowed into the session. They need to fill out a form and agree not to record and all of that.[36]

"Online Trials Cannot Reproduce the Gravitas and Power of In-Person Trials"

The final objection we want to address is that online trials cannot be as serious as in-person trials.

A trial is not just a delivery system for information; it is a sacred, public ritual space designed to create feelings of awe, social connection, and commitment to a common code of conduct and truth-finding. It is also a space where fates are decided and important decisions are made. This is why people care so much about what happens with trials, and why so many people resist moving from in-person trials, which have evolved over hundreds of years, to mediated trials. This is also why the stage and the cinema are filled with courtroom dramas (and why skeptics might say that there will never be a blockbuster movie focusing on an online trial).

36. Interview by Josh Splansky with Phil Doreau, IT Project Manager and Administrator, and Lynne Shannon, A/V Specialist, U.S. District Court for the District of Massachusetts.

Caring deeply can prevent us from seeing clearly—in this case, seeing that while mediated trials are different, they are not impossible nor are they inherently bad. They do require fresh thinking about first principles: what are the rituals of trial designed to accomplish? Having returned to first principles, we can then consider, how do we accomplish what we want to accomplish in the different environment of an online jury selection or an online trial?

The process of adopting new communication technologies follows a predictable path. First, people try to reproduce existing ways of practice using the new technology because they do not fully grasp yet both the limitations and the possibilities of this new medium. They try to do things like they have always done them. A perfect example is when movie cameras were first invented, filmmakers set up a camera in the back of a theatre and filmed a live theatre performance from a static view, as if the camera were an audience member seated in the back of the house. These films were incredibly boring. It was not until innovators like Sergei Eisenstein began experimenting with editing and creating stories in film by using its unique properties that movies as we know them started to emerge. That is the second stage—people playing with and using the new technology start to understand the unique technical possibilities of this new medium and begin to create new ways of doing things. The third stage is that new social personae begin to develop around this new medium. Movies created an entirely new type of actor, for example—one who did not need to have powerful lungs and be able to project his or her voice across a crowded theater; one who could take advantage of extremely subtle facial movements and gestures to convey character. When we look now at some of the performances of famous 19th and early 20th century actors that were captured on film, they seem overly histrionic and over-the-top. That is because we have become used to the more naturalistic performances that film and video both make possible and demand.

So what can we do to create the gravitas and presence of in-person trials in a situation where the judge and staff do not control the entire physical space? Here are some ideas:

- Use pretrial emails and communications with jurors to convey the solemnity and seriousness of the trial to jurors.
- When an online jury group assembles for voir dire, have all jurors affirm the oath together, watching each other while they do it; consider having a video of the flag waving or some other similar symbol as a screen share.

- At key points in the process, ask jurors to stand in front of their cameras and in turn, one at a time, respond to the same question, such as, "Are you ready to serve as a fair and impartial juror in this case, listening to all of the evidence with an open mind, and applying the law as I give you?" or "Do you understand and accept the law as I have presented it to you here?" This helps to create an effect similar to what happens in a room where all are physically present and recite the same oath, and a little bit like "the wave" that sports fans create in stadiums.
- Most courts already show jurors videos that teach jurors about their role and the basis of jury service. Send jurors invitations to watch these videos in videoconferencing groups with other jurors prior to the start of jury selection.
    - Consider having court staff host breakout rooms for prospective jurors to discuss what they learned in the video and ask questions.
    - Education research shows that adults remember and integrate what they learn better if they are given frequent quizzes. Have jurors complete surveys to answer questions about what they learned from the orientation videos.
- Take a cue from Hollywood. Have skilled graphic artists create "opening credits" for trial that help set the scene and mood for trial.
- Use music to convey a sense of gravity and to cue people to transitions.
- Train the technical bailiff to use his or her voice to create a sense of order, structure, and solemnity to the proceedings. (We have found that having the bailiff appear through voice only helps jurors focus and follow instructions more easily.)

These ideas by no means exhaust the possibilities. We call upon the creativity of you, our reader, and our colleagues, to create other ideas for rituals and practices to invoke the aura of justice across the wires and into the virtual courtroom.

Where to Go from Here

From a narrow, technical point of view, secure, quality online trials can be carried out. We know because we and our colleagues have been doing this for the past year. This does not mean that current systems are optimal for holding online trials.

We need videoconferencing programs designed with trials in mind, not just added on as an afterthought. In our report on our demonstration trial in June 2020, the OCP included a list of features that we felt a custom online trial platform should have; we have updated that list here in Appendix E. Trials are events with unique requirements. Technically, videoconferencing programs like Zoom, BlueJeans, and Microsoft Teams were developed to serve the needs of the business and nonprofit communities, not the courts and the trial system. They work well for meetings of equals. Trials require a more hierarchical system of control and better mechanisms for ensuring security and managing and presenting documents and other evidence to witnesses, the judge or jury, and the viewing public.

As we have discussed, we need to have a committee of experts and stakeholders come together to consider the long-term security and privacy implications of routinely broadcasting trials and develop technical solutions and court protocols for creating and storing video and audio trial records.

However, the biggest technical challenges of new technology are rarely, strictly speaking, technical. When people say that technology "doesn't work," rarely is it because a piece of equipment is broken. It is almost always because there is a mismatch between the existing way of doing things and the way the new technology forces people to work. The biggest challenges are social and cultural—incorporating new ways of doing things into existing systems—putting old wine in new bottles.

Making technology work in online trials is less about figuring out the right microphone to use and more about reimagining and redesigning the roles of courtroom staff and trial attorneys as well as the rituals of trial. As trial consultants and trial technicians, for many years, we have been playing the role of translating technology into the courtroom environment so we have a unique perspective on this issue. We hope the observations we have offered in this chapter serve as a small opening in a larger, ongoing conversation that unfolds as American court systems continue to adopt online trials as part of our justice system.

> *"Sometimes we come across technologies not because we want them, but because we need them. And then, once we need them, we want them. It's an interesting phenomenon. I can think of only a handful of times in my entire life that technology has served such a magnificent function as it is serving right now with online trials."*
>
> —Rob Rosenberg, President, Rosenberg Consulting Services, Inc.

Chapter 7

Addressing the Legal Permissibility of Online Litigation

Michael Shammas and Michael Pressman

The coronavirus pandemic uprooted life as we knew it. No institution was immune. As a result of the crisis, trials in every state (jury and bench, criminal and civil) were continued until later dates. But must they have been continued? Could they be constitutionally conducted over remote videoconferencing platforms like Zoom, Webex, or Microsoft Teams? And could they be conducted digitally absent the consent of one (or both) parties?

While only a small percentage of courts conducted online trials during the coronavirus pandemic, many attorneys conducted virtual depositions and courts held various (generally non-jury) proceedings online. While the list of proceedings occurring by videoconference was small during the early parts of the pandemic as courts tried to understand and adapt to health and safety guidelines, it was not insignificant; for example, below are some instances of trials held from the beginning of the pandemic until late spring 2020:

- The U.S. District Court for the Southern District of New York began allowing grand jurors to convene and deliberate via videoconference. (Rule 6 of the Federal Rules of Criminal Procedure does not require that a grand jury be physically located in the same room during deliberations.)
- U.S. District Judge Alison Nathan in Manhattan allowed one of the 11 jurors in the trial of an Iranian banker to deliberate by FaceTime because the juror reported feeling unwell. In light of the coronavirus, Judge Nathan stated the court was under "extraordinary circumstances" and in "untested waters."[1] After

1. Richard Gabriel, *What Online Jury Trials Could Look Like*, LAW360, Mar. 26, 2020, https://www.law360.com/articles/1257185/what-online-jury-trials-could-look-like.

being assured the juror would be secluded in an apartment, Judge Nathan stated to the juror, "You must think of yourself as present in the jury room."[2]

- In an order staying jury trials, Chief Justice of California Tani G. Cantil-Sakauye noted in passing that "[c]ourts may conduct such a trial at an earlier date, upon a finding of good cause shown or through the use of remote technology, when appropriate."[3]
- Judge Beau A. Miller of Houston, Texas, started overseeing a bench trial using videoconference technology on Wednesday, April 22, 2020.

As the pandemic progressed, more and more courts began conducting fully online criminal and civil trials. As one of this chapter's authors wrote in another paper, co-authored with Professor Christopher T. Robertson of Boston University School of Law and detailing these developments that occurred as spring progressed into summer and fall, at least three online jury trials—state and federal—occurred during this period of time:

> The nation's first pandemic-era online civil jury trial was held in Texas state court on May 18, 2020. Although the verdict was non-binding, the trial was nonetheless notable because of how relatively smoothly it went. Texas blazed another path when, in August 2020, a "virtual jury trial . . . brought about by the pandemic" was conducted by a state criminal court in Travis County, Texas. Watched by approximately 1,000 viewers, the case involved "misdemeanor charges alleging excessive speed in a construction zone," and—after using a "private virtual room to review the evidence and deliberate"—the jury returned a verdict of guilty on the speeding charge and not-guilty on the work-zone enhancement charge. A third online jury trial soon followed, this time in a federal civil court in Seattle, Washington, yielding a $1.35 million verdict. Together, the three trials—especially the

2. Ben Feuerherd, *Sick Juror Allowed to Deliberate via FaceTime amid Coronavirus*, N.Y. Post, Mar. 16, 2020, https://nypost.com/2020/03/16/sick-juror-allowed-to-deliberate-via-facetime-amid-coronavirus/.

3. News Release, California Courts, Chief Justice Issues Statewide Order Suspending Jury Trials (Mar. 23, 2020), https://newsroom.courts.ca.gov/news/chief-justice-issues-statewide-order-suspending-jury-trials.

latter two—are likely to be studied and applied in online jury trials "attempted by other courts."[4]

Importantly, trailblazing as they may have been, these trials were *not* the only online jury trials held in 2020. For example, "in *Cayla Griffin v. Albanese Enterprise, Inc.*—which has been referred to as 'the nation's first fully remote [state civil] jury trial' with a binding verdict—Florida jurors awarded the plaintiff more than $300,000 in damages for a beating suffered at the hands of bouncers in a Jacksonville, Florida club."[5] Jury selection was held entirely online, as was the one-day trial.[6] While the pandemic continued long past initial expectations, numerous courts, including many that are listed in this book, have embraced conducting full jury trials via videoconference. But significant questions arise from this new type of court proceeding: Are online civil or criminal jury trials constitutional? And what if one or both parties object? In other words, is it legally permissible for federal and state judges to conduct jury trials remotely (e.g., over videoconferencing platforms like Zoom) even without the consent of the parties?

The Question of Permissibility

In short, conducting a jury trial over a videoconferencing platform is probably both constitutional and otherwise legally permissible in the civil context. Although not definitive, research suggests that remote trials might not be constitutional in the criminal context.

The main challenge is that this is new territory for the courts. No authority seems to exist in the U.S. Constitution or in most state analogues that prohibits this type of procedure. On the other hand, no authority explicitly permits online jury trials either. Thus, in exploring whether a federal or state appellate court would or should affirm a verdict arising out of a videoconference jury trial, we must consult analogous precedents involving videoconferencing. Because courts in other contexts (civil and

4. Christopher T. Robertson & Michael Shammas, The Jury Trial Reinvented 15–17 (Boston University School of Law, Public Law Research Paper No. 21-05, 2021) (citations omitted), https://papers.ssrn.com/sol3/papers.cfm?abstract_id=3796292.

5. *Id.*

6. *See id.* at 18–19 (citing Courtroom View Network, *Florida Remote Civil Jury Trial Pilot Program: Cayla Griffin vs. Albanese Enterprise, Inc.*, https://pages.cvn.com/duval-county-florida-remote-trial-program (last visited Nov. 6, 2021)).

criminal) apply balancing tests, a similar approach would likely be used in the context of videoconference jury trials.

While due process concerns might render a digital jury trial unconstitutional in criminal cases, applying a balancing test to the public health crisis weighs in favor of allowing videoconference jury trials in civil cases. In fact, a civil jury trial by videoconferencing might not only be *permissible*; it might also be *necessary* to actualize the Seventh Amendment's protections, especially with the COVID-19 pandemic having continued longer than initially expected. The following explores the question of permissibility in both the criminal and civil contexts—and, in so doing, it explores the ways in which the issues confronted in the two contexts are both similar and different. Finally, after offering this analysis, the chapter reports on some of the initial decisions that have started to trickle in from courts across the country that have begun to confront motions and appeals challenging the use of remote proceedings.

Analysis

While one could argue that the lack of any categorical prohibition on jury trials via videoconference is itself dispositive, we explore, in what follows, ways that the question of the permissibility of videoconferencing has been addressed in contexts other than that of an entire jury trial. Courts' treatment of videoconferencing's permissibility in these contexts suggests a similar analysis would apply to entirely remote jury trials, and that such trials would be permissible in federal court and in most state courts. The following are the five main areas of analysis in examining the permissibility of trials using videoconferencing technology.

Public Trials

American justice is public justice. As the U.S. Supreme Court wrote (regarding criminal cases), public trials are "essential" since "the presence of interested spectators . . . keep [the defendant's] triers keenly alive to a sense of their responsibility and to the importance of their functions."[7] Although the Constitution does not require it, courts have clarified that civil jury trials, too, should be just as public as criminal ones. The California Supreme Court is representative: "We believe that the public has

7. Waller v. Georgia, 467 U.S. 39, 46 (1984) (internal citations omitted).

an interest, in *all* civil cases, in observing and assessing the performance of its public judicial system."[8]

From a technical standpoint, there is no practical reason why online jury trials cannot be public. Many of the remote trials that were conducted during the pandemic posted YouTube links where the public could view the online proceedings. Most Americans have adapted to videoconferencing technology in their work, their children's schooling, and their social interactions during the pandemic, so there is no barrier to public knowledge of or ability to access court proceedings. Numerous states have also been making digital hearings public—including to citizens without internet—by setting up rooms in courthouses where citizens can watch online proceedings.[9] Accordingly, public libraries, courthouses, and other government or community buildings could ensure that all Americans could have the opportunity to watch trials. However, if courts allow digital jury trials, the risk of certain pitfalls—such as the possibility of someone recording the proceedings—must be clearly communicated to the parties in advance.

The Seventh Amendment and State Equivalents[10]

The Seventh Amendment guarantees every American the right to trial by jury in most civil cases. While the substance of the amendment has never been incorporated, most states comply with the jury trial requirement anyway. As Justin Sarno and Jayme Long write in Law360:

> According to the U.S. Supreme Court in *Simler v. Connor*, "[t]he federal policy favoring jury trials is of historic and continuing strength." In fact, according to the Federal Rules of Civil Procedure, Rule 38, "[t]he right of trial by jury as declared by the Seventh Amendment to the Constitution—or as provided by a federal statute—is preserved to the parties *inviolate*."[11]

8. NBC Subsidiary (KNBC-TV), Inc. v. Superior Court, 20 Cal. 4th 1178, 1210 (1999).

9. *See, e.g.*, Jamiles Lartey, *The Judge Will See You on Zoom, but the Public Is Mostly Left Out*, Marshall Project, Apr. 13, 2020, https://www.themarshallproject.org/2020/04/13/the-judge-will-see-you-on-zoom-but-the-public-is-mostly-left-out.

10. The Supreme Court has said multiple times that states do not have to provide jury trials in civil cases. Most states nonetheless guarantee the right to a civil jury trial, and (regardless of whether or not they do) they must do so in cases decided under federal law.

11. Justin Sarno & Jayme Long, *Social Distancing and Right to Jury Trial Must Be Reconciled*, Law360, Apr. 12, 2020, https://www.law360.com/articles/1261476/social-distancing-and-right-to-jury-trial-must-be-reconciled.

Given the respect Americans grant the right to a jury trial, in the absence of juries, the judicial system could experience a huge backlog of civil cases, eliminate a civic institution that the American public relies on, affect the willingness of plaintiffs to sue, and impact public policy.

So the constitutional question is not just academic but practical. Indeed, one could argue that the Constitution *requires* judges to find a way to actualize the Seventh Amendment by restoring trial by jury—even, if necessary, by videoconference.

Recognizing the necessity of a functioning judiciary, even in a pandemic, the federal government has brainstormed contingency plans aimed at keeping courts open during a time of social distancing. The most recent attempt stems from the George W. Bush administration. In March 2007, the U.S. Department of Justice Office of Justice Programs worked with American University's School of Public Affairs and the Bureau of Justice Assistance to release a publication entitled "Guidelines for Pandemic Emergency Preparedness Planning: A Road Map for Courts."

While the road map is helpful, it is more useful in a logistical sense than in a legal sense. Guidance on constitutional questions regarding the Seventh Amendment (or, indeed, the Sixth Amendment) is nowhere to be seen. However, this contingency plan, surely vetted by the administration's counsel to ensure the continued operation of courts during a potential pandemic, also suggests that online jury trials are permissible.

The Confrontation Clause

Several cases address rights provided by the Confrontation Clause.[12]

---

12. *See, e.g.*, Illinois v. Allen, 397 U.S. 337, 338 (1970) ("One of the most basic of the rights guaranteed by the Confrontation Clause is the accused's right to be present in the courtroom at every stage of his trial."); Kentucky v. Stincer, 482 U.S. 730, 745 (1987) ("The Court has assumed that [the defendant] has a due process right . . . to be present at any stage of the criminal proceeding that is critical to its outcome if his presence would contribute to the fairness of the procedure."); United States v. Gordon, 829 F.2d 119, 123 (D.C. Cir. 1987) (maintaining that Federal Rule of Criminal Procedure 43(a), affording a defendant's right to be present during jury empanelment, "embodies the protections afforded by the Sixth Amendment Confrontation Clause, the due process guarantee of the fifth and fourteenth amendments, and the common law right of presence"); United States v. Novaton, 271 F.3d 968, 998 (11th Cir. 2001) (holding that a criminal defendant's constitutional right to be present under the Confrontation Clause and the Fifth Amendment's Due Process Clause was violated by conducting portions of his trial when the defendant was absent due to illness and when he protested).

Presence

Criminal cases involving witness testimony against the accused via closed-circuit television (CCTV) provide further guidance.[13] In *Craig*, the Supreme Court held that the Confrontation Clause does not categorically forbid child abuse victims from testifying against defendants via one-way CCTV.[14] In disposing of the key constitutional issue—the presence of the defendant before his accuser—Justice O'Connor wrote that the Confrontation Clause merely embodies a preference (not a requirement) for in-person confrontation, which can accordingly be limited to satisfy sufficiently important interests.[15]

Analogizing from the use of CCTV to the use of videoconference software, it seems that the Confrontation Clause's presence requirement would be satisfied if the testimony is (1) reliable and (2) if a specific necessity has been shown for videoconferencing.[16] More broadly, *Craig* suggests that traditional constitutional rights (like a defendant's presence before his or her accuser) can be satisfied or modified if videoconferencing satisfies other sufficiently important interests. The challenge posed by COVID-19 (i.e., the continued operation of the judiciary at a time when in-person jury trials could be deadly) seemingly rises to a sufficiently important level—one "sufficiently important to justify the use of a special procedure" such as videoconferencing.[17]

Regarding civil cases, the question of presence is substantially easier. In civil trials, courts decide whether to allow videoconferencing by employing a balancing-of-interests test. For example, in cases brought by prisoners, the Supreme Court has held that the constitutional right of meaningful access to the courts must be balanced against the state's interest in keeping prisoners confined (due to logistical and safety concerns).[18]

Observation

Another takeaway from Sixth Amendment Confrontation Clause jurisprudence is that whether a testifying witness can be *observed* is crucial.

---

13. *See, e.g.*, Maryland v. Craig, 497 U.S. 836, 857 (1990).
14. *Id.*
15. *Id.*
16. *Id.* at 855 ("[W]e hold that, if the State makes an adequate showing of necessity, the state interest in protecting child witnesses from the trauma of testifying in a child abuse case is sufficiently important to justify the use of a special [(e.g., videoconferencing)] procedure.").
17. *Id.*
18. Bounds v. Smith, 430 U.S. 817, 821 (1977).

Any videoconferencing platform that does not allow for easy observation would be impermissible.

Again, though the Confrontation Clause does not apply to civil jury trials, *Craig*'s emphasis on observation is a guidepost because allowing jurors to assess witness credibility is as important in weighing evidence in a civil jury trial as it is in a criminal jury trial.[19]

Interestingly, platforms like Zoom may in fact make credibility determinations *easier* because one can see every participant at a glance.[20] Anecdotal observations by jurors and judges contained in other chapters of this book suggest that jurors do indeed have this experience—the witness is full face in front of them—thus they can see the witness's expressions more clearly.

Finally—and importantly—it should be noted that in light of the diversity of state constitutions, analogues of the Confrontation Clause are inapplicable in some states.[21]

---

19. The Court found:

> In sum, we conclude that . . . the Confrontation Clause does not prohibit use of a procedure that, despite the absence of face-to-face confrontation, ensures the reliability of the evidence by subjecting it to rigorous adversarial testing and thereby preserves the essence of effective confrontation. Because there is no dispute that the child witnesses in this case testified under oath, were subject to full cross-examination, *and were able to be observed by the judge, jury, and defendant as they testified*, we conclude that, to the extent that a proper finding of necessity has been made, the admission of such testimony would be consonant with the Confrontation Clause.

Craig, 497 U.S. at 857 (emphasis added).

20. Some commentators have worried that jurors will pay less attention in an electronically conducted trial than in a "real" trial. Others have worried that—especially in the criminal context—jurors or judges may be more willing to give out harsh penalties (to either party) if they are not in the same room as the parties. These concerns clearly have merit.

21. *See* People v. Fitzpatrick, 158 Ill. 2d 360, 365 (1994) ("[W]e conclude that the confrontation clause of the Illinois Constitution provides that a defendant is entitled to a face-to-face confrontation with a witness."); Commonwealth v. Bergstrom, 524 N.E.2d 366, 371–72 (1988) ("The plain meaning of assuring a defendant the right 'to meet the witnesses against him face to face' is that the accused shall not be tried without the presence . . . of both himself and the witnesses testifying against him."); *but see* People v. Phillips, 315 P.3d 136, 152 (Colo. Ct. App. 2012) (citing *Craig* and holding that neither the Colorado nor federal constitution render CCTV unconstitutional).

Arraignments

Although the lack of physical presence at arraignments might not be unconstitutional,[22] the Federal Rules of Criminal Procedure appear to require physical presence.[23] Even in this context, however, courts have held that videoconferencing may satisfy "presence" for arraignments when necessary.[24]

The possibility of videoconferencing satisfying "presence" if necessity is shown is yet another instance of a context that weighs in favor of permitting remote jury trials during a pandemic or other national emergency. As *Maryland v. Craig* was decided in 1990, *United States v. Lawrence* in 2001, *People v. Fitzpatrick* in 1994, and *Commonwealth v. Bergstrom* in 2012, before the development of our current videoconferencing technology, future appellate and supreme courts may have to define whether the terms "face to face" or "presence" are satisfied in a remote setting, given the current state of technology.

Federal Rules of Civil Procedure and State Analogues

Federal Rule of Civil Procedure 43(a) has been used in the civil and criminal context to help courts determine the permissibility (or lack thereof) of CCTV.[25] Rule 43(a) allows videoconferencing for "good cause in compelling circumstances with appropriate safeguards."[26] Further:

> This standard considers the relative cost of transporting the witnesses to court in comparison to the cost for videoconferencing, the ability of the court to use subpoena power over the proposed witnesses, whether the alleged criminal incident occurred overseas, and whether United States officials will be available to swear in witnesses at their location.[27]

---

22. Valenzuela-Gonzalez v. U.S. Dist. Court for Dist. of Ariz., 915 F.2d 1276, 1279 (9th Cir. 1990).

23. *Id.*; FED. R. CRIM. P. 43(a).

24. *See* United States v. Lawrence, 248 F.3d 300, 305 (4th Cir. 2001).

25. United States v. Guild, 2008 WL 191184, at *3 (E.D. Va. Jan. 17, 2008) ("Recognizing both the importance of live testimony in a criminal trial and the fact that the Confrontation Clause is not implicated by this testimony, the Court will use Federal Rule of Civil Procedure 43(a) as the threshold showing for the use of videoconferencing in this instance.").

26. FED. R. CIV. P. 43(a).

27. DANIEL DEVOE & SARITA FRATTAROLI, MASSACHUSETTS SOCIAL LAW LIBRARY, VIDEOCONFERENCING IN THE COURTROOM: BENEFITS, CONCERNS, AND HOW TO MOVE FORWARD (2009).

Considering Rule 43(a) along with various state rules of civil procedure, three things are clear: First, both constitutional *and* procedural questions will arise as videoconferencing is used more and more in courts. Constitutional objections to videoconferenced trials will of course be more common in criminal cases, but procedural concerns are likely to pervade both the civil and criminal contexts. Accordingly, the Federal Rules of Civil Procedure and state analogues must be updated to help lawyers and judges smoothly transition into a uniform procedure for trial by videoconference.

Second, the state rules closely mirror Rule 43(a), so cases analyzing Rule 43(a) will be applicable at the state level as well. Maine's rule is representative:

> In every trial, the testimony of witnesses shall be taken in open court, unless a statute, these rules or the Rules of Evidence provide otherwise. *The court may, on its own motion or for good cause shown upon appropriate safeguards, permit presentation of testimony in open court by contemporaneous transmission from a different location.* All evidence shall be admitted which is admissible under the statutes of this state, or under the rules of evidence applied in the courts of this state.[28]

Third, and as apparent from the Maine rule, most of the state rules of civil procedure (like the federal rule) require a balancing test. The question of whether trial by videoconference is permissible may therefore differ by state. The balance of interests may make trial via videoconference necessary in New York; it may not do the same in North Dakota.

Recent Developments

As this book goes to print, courts across the country that have conducted proceedings remotely have begun to confront legal challenges to the remote nature of these proceedings—raised by parties to the cases in which judges conducted remote proceedings. Accordingly, courts have begun to issue decisions addressing these challenges. At this moment, these decisions are few and far between, but they are likely to grow in number as time passes and more parties appeal, and as more proceedings are conducted remotely. Not only will many more types of remote proceedings be addressed by courts going forward, but, so too, the types

28. ME. R. CIV. P. 43 (emphasis added).

of proceedings addressed in the rulings that have been issued to date may be addressed by other courts (including higher courts), and there is of course the possibility that these other courts (including higher courts) will come to conclusions that are similar or different from the conclusions arrived at in these early decisions. In what follows, we flag a few recent decisions that are among the first to address the permissibility of remote proceedings.

Western District of Washington

Judge Marsha J. Pechman, U.S. district judge for the Western District of Washington, recently issued an opinion in *Goldstine v. FedEx Freight, Inc.*,[29] a civil case for which she had conducted a fully remote eight-day jury trial. In her opinion, she denies post-trial motions from both parties and also provides other findings and conclusions. Relevant for the purposes of this chapter is the fact that one of the arguments in the defendant's post-trial motion is that the remote trial violates the Seventh Amendment, Rule 43(a) of the Federal Rules of Civil Procedure, and the court's "General Order No. 15-20." Judge Pechman denied both of the post-trial motions in full, and, accordingly, she denied the defendant's arguments that the remote trial violated the defendant's rights. Judge Pechman's opinion constitutes a valuable blueprint for courts addressing these issues.

Supreme Judicial Court of Massachusetts

The Supreme Judicial Court of Massachusetts recently confronted the question of whether a remote suppression hearing violated a criminal defendant's due process rights, and it concluded that it did not.[30] Following previous cases, the court stated that due process was a "flexible" concept that might vary with the context. Addressing the various specific rights of the criminal defendant—many of which have been discussed in this chapter—the court concluded that remote proceedings were not a per se violation of these rights. Among other things, the court held that the defendant's right to be present was safeguarded effectively as long as the remote hearing allowed the defendant "to listen to the evidence, adequately observe the witnesses who testify at the hearing, and privately consult with his attorney at any time,"[31] and that the right to a public trial was satisfied

29. 2021 U.S. Dist. LEXIS 46478 (W.D. Wash. Mar. 11, 2021).
30. Vazquez Diaz v. Commonwealth, 487 Mass. 336 (2021).
31. *Id.* at 342.

by the superior court's standing order allowing public access to the remote proceedings via telephone or remote platforms such as Zoom.[32]

Supreme Court of New Jersey

The Supreme Court of New Jersey has also recently been confronting questions regarding the permissibility of remote proceedings. In one recent decision, the Supreme Court of New Jersey held that remote grand jury proceedings during the pandemic were consistent with constitutional protections.[33] Second, the Supreme Court of New Jersey will soon be deciding a different question regarding remote proceedings when it issues an opinion on an appeal in a case that is currently pending.[34] In *New Jersey v. Dangcil*—an appeal by a criminal defendant who was prosecuted for attempted arson in the fall of 2020—the New Jersey Supreme Court will be deciding whether it is unconstitutional to conduct remote jury selection during the current pandemic. Although the trial itself was held in person, the jury selection process took place remotely in October 2020. The defendant initially brought an emergency challenge to the use of remote jury selection at that time, but the emergency challenge was denied vis-à-vis the emergency basis; the court said, however, that the defendant could challenge the use of remote jury selection after the trial had concluded. With the trial now having concluded, however, the defendant reissued his challenge, and the court stated that "[d]efendant's motion for direct certification is granted, limited to defendant's challenge to the hybrid virtual/in-person jury selection procedure." As this book goes to print, the case remains pending before the New Jersey Supreme Court.

Conclusion

Ultimately, in exploring whether courts can conduct entire jury trials by videoconference, we tread on new terrain. While the Supreme Court of Missouri has remanded three criminal cases stating the use of video testimony violated the criminal defendants' rights to confront their accusers, a Minnesota state court of appeal has found that use of live video testimony did not violate those same constitutional rights.[35] Ultimately, the U.S. Supreme Court will have to resolve these conflicting rulings.

32. *Id.* at 353.
33. State v. Vega-Larregui, 246 N.J. 94 (N.J. 2021).
34. New Jersey v. Dangcil, No. 084990 (N.J.).
35. Missouri v. Smith, No. SC99086 (Mo. Jan. 11, 2022); Minnesota v. Tate, No. A21-0359 (Minn. Jan. 3, 2022).

Any decision as to whether civil or criminal jury trials held via videoconference are permissible would and should likely undergo a balancing approach. With its weighty constitutional protections, the criminal context will likely be where videoconferencing remains most controversial. In the civil context, however, jury trials held via videoconference may well be permissible, even when one of the parties does not consent. Although a videoconference jury trial might not be ideal for some cases, any drawbacks (such as the potential for reduced juror attention) can simply be added into the balance of costs and benefits. As courts and society on the whole become more accustomed to Zoom, Webex, Microsoft Teams, and similar platforms, these drawbacks will likely be mitigated.

Chapter 8

Preserving the Representativeness of Jury Pools in Online Trials

Sarah Murray, with assistance from Marc King

> *"Most people don't think about how a different way of doing things might make things better—they only think about how it might make them worse."*
>
> —Judge Emily Miskel, 470th District Court of Collin County, Texas

Introduction

Even before the COVID-19 pandemic, many Americans were concerned about the growing "digital divide" and relative lack of broadband access by certain social classes and groups, including rural Americans, low-income Americans, Native Americans, and Black, Hispanic, and Latino Americans. As the pandemic thrust us into a new world of online hearings and trials and computer and internet access has more directly determined citizens' access to justice, this issue has received heightened attention and concern from lawyers, judges, and others in the justice system. However, often, more heat than light has been cast in public discussion of this issue, with many simply asserting that online jury pools cannot be representative or expressing concerns and fears without appeal to actual data and experience.

My goal is this chapter is to answer four questions related to representation access issues and jury trials in the United States:

1. What do we know about the representativeness of in-person jury pools prior to the COVID-19 pandemic? This provides an

important baseline to any discussions of how online trials might impact jury pool representativeness.

2. What do the data show about Americans' current access to computing technology and internet broadband services generally?
3. What have judges and court administrators who have been conducting online jury trials during the pandemic observed about online jury pools and online jury service compared to in-person jury pools and jury service, in terms of both possibilities and problems?
4. What are the best practices for ensuring that all citizens have an equal ability to successfully participate in an online jury selection or serve on an online jury?

While this chapter focuses specifically on the issue of representativeness in online jury pools, it is worth noting that jury trials are rare in both state and federal courts today, and most Americans encounter the justice system through family law courts, probate courts, traffic courts, and drug courts—not through jury service. Access to justice does not rest on a representative group of citizens being able to serve as jurors, but more broadly on citizens from all walks of life and socioeconomic circumstances being able to attend court proceedings that directly affect their lives and communities. All of the judges and court administrators we interviewed or heard from discussed this issue and commented on the promise of online trials to allow jurors, litigants, family members, and the public from all walks of life to more easily attend court proceedings.[1]

The Promise and Realities of Representation Pre-COVID

The most basic role of trial by jury has been to protect members of the community from oppression by the government. The Constitution itself

1. Judge Emily Miskel says:

> The last time I looked at the numbers, about 20% of my cases were resolved through a bench trial and .2% of my cases were resolved through a jury trial. One time, I went two years with zero jury trials requested. What I have seen as much as representativeness involves family law. About 50% of cases handled in Texas are family law cases. That is how most people interact with the court system.

Interview by Marc King with Judge Emily Miskel, 470th District Court, Texas (Apr. 9, 2021) [hereinafter Interview with Judge Miskel].

does not guarantee a jury pool that is representative of the community; it speaks only of "an impartial jury of the state and district wherein the crime shall have been committed."[2]

What Is a "Representative" Jury Pool?

Traditionally, however, in American law, "an impartial jury" has been interpreted to mean a jury of one's peers, which in turn over time has come to be understood by the U.S. Supreme Court as a representative group of citizens who live in the venue in which the case is being tried. In *Taylor v. Louisiana*,[3] the Court wrote, "The selection of a petit jury from a representative cross section of the community is an essential component of the Sixth Amendment right to a jury trial."

According to the Supreme Court's decision in the criminal case *Duren v. Missouri*,[4] to establish a prima facie violation of the fair-cross section requirement, the defendant must show, among other things, that the group alleged to be excluded is a "distinctive" group in the community.[5] The Court has not defined the term "distinctive"; in practice, it has most often been used to protect against exclusion of citizens based on race, ethnicity, and gender in criminal cases; the focus on these categories comes in great part from the historic exclusion of racial minorities and women from government.

However, from the point of view of jury decision-making, representativeness is not just an issue in criminal cases or in cases with defendants who come from marginalized or minority groups, and race, ethnicity, and gender are not the only sociological categories that may be salient to ensuring a representative mix of attitudes and experiences on a jury. In both civil and criminal cases, for example, the lack of representation from groups like hourly paid workers, management and executive

---

2. The entire text of the Sixth Amendment reads:

> In all criminal prosecutions, the accused shall enjoy the right to a speedy and public trial, by an impartial jury of the state and district wherein the crime shall have been committed, which district shall have been previously ascertained by law, and to be informed of the nature and cause of the accusation; to be confronted with the witnesses against him; to have compulsory process for obtaining witnesses in his favor, and to have the assistance of counsel for his defense.

3. 419 U.S. 522, 528 (1975).
4. 439 U.S. 357, 364 (1979).
5. Strengthening the Sixth, *Impartial & Representative Juries*, https://www.strengthenthesixth.org/focus/Impartial-Representative-Jury (last visited Nov. 6, 2021).

workers, people with disabilities, or younger people could be problematic, depending on the parties and issues at stake—although these are not typically the categories that have been considered by attorneys or courts in deciding if a jury is representative.

Recognizing that many factors shape the composition of the final seated *jury*, the law does not require that a seated jury be a representative cross-section of the community, only that the broader jury *pool* be representative. Let us say, for example, that a jury pool of 60 people is called for jury selection in a month-long trial. If that pool of 60 fairly reflects the demographics of the venue in characteristics such as race, gender, age, education, and economic status, that is sufficient to meet the constitutional requirements, even if the actual seated jury does not include any low-income citizens because all were dismissed for hardship because they could not afford to serve on such a lengthy trial. Ensuring that the pool of citizens summoned for jury duty is representative of the community is only the first step in making jury pools and actual seated juries more representative, however, as many obstacles stand in the way of citizens who are called for jury duty actually showing up to court and serving as jurors. As Judge Emily Miskel from Texas expressed when discussing the difference between in-person and online trials, "The dirty secret is that in person trials exclude a lot of people."[6] Indeed, understanding what jury pools and jury service looked like with in-person trials before COVID—what social scientists would call understanding the baseline conditions—is vital to evaluating the impacts on representativeness of holding jury trials online.

Barriers to Jury Service

Around the country, the number of companies that pay for a week or more of jury service has declined significantly over my 23 years as a trial consultant, and many more citizens work as subcontractors or in "gig economy" jobs. Citizens who are hourly paid workers and are not paid for jury duty often cannot afford to take off work; even many salaried workers work for companies that do not pay for jury duty or pay for only three days of service or less and cannot afford to serve the length of most cases that go to trial.[7] Citizens in this situation often simply ignore jury summons, sometimes at the direction of their employer. I know this from

6. Interview with Judge Miskel, *supra* note 1.

7. Currently, only eight states out of 50 require employers to pay employees who are on jury duty: Alabama, Colorado, Connecticut, Louisiana, Massachusetts, Nebraska, New York, and Tennessee.

personal experience—I worked as a grill cook while in graduate school. When I received a jury summons, my boss told me to throw it in the trash because if I were chosen to serve on a jury, the restaurant would have to close down as they had no one else who could work my shifts, as the job required special skills and experience.

Even citizens who want to do their duty are often without a car or reliable transportation. Judge Miskel commented, "Our system already excludes people who lack the technology of a car. We don't send a taxi for every person who lacks a car."[8] Many may be unable to afford necessary childcare, or may have other family and care responsibilities that make it difficult or impossible to take the time required to go to court. In addition, some citizens are afraid of going to court for a variety of reasons.

Many states also have laws that exclude individuals who have been convicted of felonies or incarcerated from serving on juries. Given that there is a significant racial skew in who is arrested and convicted of crimes across the United States, this creates a baseline pool of eligible jurors that is similarly skewed. Recognition of this has led to a political movement to restore the rights of felons who have served their time to vote and to serve on juries; California passed such a law in 2019.[9] However, still about half of U.S. states ban jury service by ex-felons. Since Black Americans are more likely to be convicted felons, this can significantly impact racial representation in jury pools.

Pre-COVID Juror Show Rates Vary Widely in Both State and Federal Courts

Pre-COVID judges and courts around the country reported show rates in response to jury summons that varied widely. For example, in May 2015, a citizens group, California Citizens Against Lawsuit Abuse (CALA), published a report on jury show rates based on analysis of data provided through public record requests by the 15 most populous California counties.[10] They found that the rate of jurors who failed to appear for jury service varied widely among the counties, ranging from a high of 45.66 percent for Ventura County to a low of 3.90 percent

8. Interview with Judge Miskel, *supra* note 1.

9. Debra Cassens Weiss, *New California Law Allows Felons Who Served Their Time to Serve on Juries*, A.B.A. J., Oct. 11, 2019, https://www.abajournal.com/news/article/new-california-law-allows-felons-who-served-their-time-to-serve-on-juries.

10. CALA, Do Californians Answer the Call to Serve on a Jury? A Report on California Rates of Jury Service Participation (2015), https://californiacala.org/reports-1/do-californians-answer-the-call-to-serve-on-a-jury-a-report-on-california-rates-of-jury-service-participation.

for San Joaquin County, with a weighted average of 20.43 percent who failed to appear. In May 2017, the Associated Press published a story on problems with juror show rates that cited a national average no-show rate of 9 percent in state cases, but said that some courts have no-show rates as high as 50 percent.[11] In that same story, the article states that in Multnomah County Circuit Court in Oregon, a typical show rate is 25 percent, with 75 percent of those called failing to appear. According to Judge Miskel, who serves in Collin County, Texas (part of the Dallas-Fort Worth metro area), typical show rates in her courtroom were 40 percent; based on conversations with her colleagues, typical show rates in Houston courtrooms were 18 percent.[12] David Slayton, statewide administrator of the Texas court system, cites a pre-pandemic statewide juror show rate of about 40 percent.[13]

While jurors are required by law to show up when called for jury duty, in practice most jurisdictions do not enforce the law and citizens who do not show up to jury duty are rarely punished.

Courts typically do not analyze the demographics of who shows up for jury duty versus those who do not to see how these show rates affect the representativeness of the pool; most courts, state and federal, do not collect the data necessary to perform this kind of analysis. In general, the collection of data on jury pools and trials is scattered and fragmentary in the United States, making it difficult to make empirically sound national analyses of representativeness.[14] The only state-level analysis of jury pool demographics and barriers to service of which I am aware is a report prepared based on survey data collected in 2016–2017 from courts

11. Maxine Bernstein, *Judges Cracking Down on People Who Snub Jury Duty*, AP News, May 21, 2017, https://apnews.com/article/62b279c38615469fb9bee505c9c66ff5.

12. Interview with Judge Miskel, *supra* note 1.

13. David Slayton, State of Texas Office of Court Administration, Jury Trials during the COVID-19 Pandemic: Observations and Recommendations (2020).

14. The National Center for State Courts and the Conference of State Court Administrators run a joint Court Statistics Project that collects standardized data on caseloads from the 50 states, the District of Columbia, and Puerto Rico, but there is no comparable project to collect data on juries and jury pools. It would be a significant undertaking, but it is an area ripe for data collection, perhaps as a joint venture between local universities and courts.

in Washington state as part of a state government initiative to address diversity issues in juries.[15] The authors report their

> [r]esults show that Black, Native/Indigenous, People of Color (BIPOC), including Asian Americans and Hispanic/Latinx Americans, are underrepresented in nearly all Washington jury pools. Findings also indicate that BIPOC women were underrepresented in all courts within this analysis. Further findings point towards differences by race and sexual orientation and may indicate underrepresentation among all racial categories.[16]

As a trial consultant who has picked hundreds of juries in my career, I have tracked the demographics of actual jury panels in hundreds of state and federal venues around the country since 1999 and routinely compared them to the general census demographics. I have had colleagues working with me do the same. Our data matches the results of the Washington state data, showing that the pool of jurors who show up to court and sit through jury selection in complex civil and criminal cases[17] is typically skewed relative to the general population.[18] These jury pools typically contain fewer working-class people; fewer high school educated people; fewer hourly wage workers; and fewer people of color than the census statistics would suggest. They also contain fewer young people, as young people are often juggling school and work and cannot take the time needed to attend an in-person trial. In other words, even before COVID-19 and the pressure to hold online trials, for in-person trials, jury pools have tended to be older, whiter, better educated, and better employed at stable, long-term jobs than average in their community.

The more anecdotal observations of Texas Judge Roy Ferguson (as relayed to us by his colleague Judge Miskel) match the above analysis. He presides over a five-county area in West Texas with a majority Hispanic and Latino population. Prior to COVID and moving to online trials, he says that his in-person jury pools and juries comprised largely retired, white citizens—certainly not representative of the citizens of his district.[19]

---

15. Peter Collins & Brooke Miller Gialopsos, *Answering the Call: An Analysis of Jury Pool Representation in Washington State*, 22 Criminology Crim. Just. L. & Soc'y 36 (2021).

16. *Id.*

17. Complex cases are cases that usually take two weeks or more to try.

18. Note that the general population includes noncitizens and citizens whose English may be inadequate to serve on a jury.

19. Interview with Judge Miskel, *supra* note 1.

While I have not conducted a formal analysis of the reasons for this skew, based on my observations and general reading of the literature, I conclude that this skew results from two factors: (1) jury summons are more likely to be successfully delivered to people who have long-term residential and financial stability, and (2) people who are better off, are paid for jury duty or can afford to take time off, are more trusting of government, and feel more invested in the status quo, are more likely to respond to a jury summons. Data published by the Pew Research Center in a 2017 report on attitudes towards voting and jury duty support this analysis, as their data shows that younger people, racial and ethnic minorities, and people without a college education are less likely to believe that jury duty is connected to good citizenship.[20] Legal scholars Paula Hannaford-Agor and Nicole L. Waters also report that "[f]ailure-to-appear rates and excusal rates are . . . highly correlated with socioeconomic status. . . ."[21]

Summary

In sum, data and experience show that, pre-COVID, the show rates of citizens called for in-person jury service varied considerably by venue and were significantly impacted by socioeconomic status, race, and ethnicity. As a result, the pre-COVID jury pools in most venues suffered significant problems with representativeness unrelated to access to computer and internet technology.

What Is the Current State of Broadband and Computer Access in the United States?

To participate successfully in an online jury trial, a juror requires:

- Access to a sufficiently capable computing device
- The availability of broadband internet service in the juror's community

20. Pew Research Center, Public Supports Aim of Making It "Easy" for All Citizens to Vote (2017), https://www.pewresearch.org/politics/2017/06/28/public-supports-aim-of-making-it-easy-for-all-citizens-to-vote/. Note that while the press release focuses on voting issues, the top-line results issued included data on a question about jury service and citizenship. *See also* John Gramlich, *Jury Duty Is Rare, but Most Americans See It as Part of Good Citizenship*, Pew Rsch. Center, Aug. 24, 2017, https://www.pewresearch.org/fact-tank/2017/08/24/jury-duty-is-rare-but-most-americans-see-it-as-part-of-good-citizenship/.

21. Paula Hannaford-Agor & Nicole L. Waters, *Safe Harbors from Fair-Cross-Section Challenges? The Practical Limitations of Measuring Representation in the Jury Pool*, 8(4) J. Empirical Legal Stud. 768 (2011).

- A temporary or ongoing subscription to a broadband internet service with sufficient speed and minutes to allow for smooth upload and download of audio and video

Available data suggests that, currently, a large majority of American households meet all three of these criteria; however, the data also shows disparities in access based primarily on three factors:

- Age, with older citizens more likely to be without internet access or access to a computing device
- Geographical location and whether a citizen lives in a rural or urban area
- Socioeconomic status

The data also shows that Black and Hispanic Americans, particularly low-income Black and Hispanic Americans, are more likely to have only a smartphone, a device that is generally inadequate to participate fully in an online trial (although it may be sufficient to participate in online jury selection).

In addition, the data shows that access to both computers and internet broadband service shows significant regional variation.

I present and discuss this data in more detail below. Before engaging in that discussion, however, I think it is important to add some nuance to the discussion of access. When considering representativeness issues, people often talk as if access to technology is a "yes or no" issue. They also rarely consider that different tasks require different levels of technology and access, and this is not a trivial point when considering the resources that might be required to ensure access to online governmental functions.

In my observation, for some, concerns about representativeness in jury pools arose or were amplified during the pandemic because many public school districts moved to online education and reports came in from around the country that children from lower-income households were facing access problems. The technological challenges that students from low-income households face in reliably connecting to school are real and represent a serious social problem—one that both public and private organizations recognize and are seeking to address in various ways. However, the nature of jury service is quite different from attending school. Jury service is occasional and requires much less time and access than daily school attendance. Solutions that might not work for students who need to attend school daily from home may well work for citizens who only need to appear for jury selection for a few hours or a few days.

Access to "Computers" in the Home

The U.S. Census began collecting data on computer access and use in the United States in 1984 and internet access in 1997, dates that correspond to the beginnings of widespread adoption of each of these technologies.[22] A "computer," per the census, includes desktop computers, laptop computers, tablets, and smartphones. U.S. Census data shows that the percentage of American households that have access to a computer in the home steadily increased between 1984 (when data on at-home computer access first began to be collected) and 2016, with the percentage of Americans with access to computers in 2016 at about 90 percent.[23]

However, as anyone who uses computing devices knows, there can be significant differences in capability between desktop and laptop computers, tablets, and smartphones. When holding trials and courtroom proceedings online, having the larger screen real estate of a desktop, laptop, or tablet computer is often critical to successful engagement and ability to view and respond to all of the evidence. Census data shows that lower-income households, Black households, and Hispanic households are more likely to be smartphone-only households; this corresponds to my experience working with mock juror groups around the country when conducting online research.[24] A recent survey by the Pew Research Center also found:

> With fewer options for online access at their disposal, many lower-income Americans are relying more on smartphones. As of early 2019, 26% of adults living in households earning less than $30,000 a year are "smartphone-dependent" internet users—meaning they own a smartphone but do not have broadband internet at home. This represents a substantial increase

---

22. The Current Population Survey (CPS), run by the Census Bureau and the U.S. Bureau of Labor Statistics, began collecting data about computer use in 1984 and internet use in 1997. In 2013, the American Community Survey (ACS) also began collecting data on those topics as mandated by the 2008 Broadband Data Improvement Act. Both of these surveys are based on probability sampling rather than the direct counts of the census. The CPS relies on a sample of about 60,000 occupied households. U.S. Census Bureau, *Current Population Survey (CPS)—Methodology*, https://www.census.gov/programs-surveys/cps/technical-documentation/methodology.html (last revised Oct. 8, 2021). The ACS has a larger sample of more than two million households for 2019. U.S. Census Bureau, *American Community Survey—Sample Size*, https://www.census.gov/acs/www/methodology/sample-size-and-data-quality/sample-size/index.php (last visited Nov. 6, 2021).

23. Camille Ryan, U.S. Census Bureau, Computer and Internet Use in the United States: 2016 (2018), https://www.census.gov/content/dam/Census/library/publications/2018/acs/ACS-39.pdf.

24. *Id.*

from 12% in 2013. In contrast, only 5% of those living in households earning $100,000 or more fall into this category in 2019.[25]

This same Pew Research study found that roughly three in ten adults with household incomes below $30,000 a year do not own a smartphone, and more than four in ten do not have home broadband services. By comparison, virtually all adults in households earning $100,000 or more have access to both of these technologies, often in addition to access to laptop, desktop, and/or tablet computers.

What is not clear from this data is what percentage of those who live in low-income households without access to smartphones or other computer technology may be residents who are not citizens and therefore do not qualify for jury service.

Broadband Internet Access

The Federal Communications Commission (FCC) is the federal agency responsible for managing the nation's communications networks and ensuring equal access for all Americans. In April 2020, the FCC issued its annual broadband deployment report summarizing the state of broadband access, stating in the Introduction:

> The vast majority of Americans—surpassing 85%—now have access to fixed terrestrial broadband service at 250/25 Mbps [megabits per second], a 47% increase since 2017. Over the same period, the number of Americans living in rural areas with access to such service increased by 85%. This progress has been fueled in part by an approximately $80 billion investment in network infrastructure in 2018, the highest annual amount in at least the last decade. In 2019 alone, fiber broadband networks became available to roughly 6.5 million additional unique homes, the largest one-year increase ever, with smaller providers accounting for 25% of these new fiber connections. AT&T, Sprint, T-Mobile, and Verizon are also rapidly expanding their 5G capability, with 5G networks in aggregate now covering the majority of the country's population, especially in urban areas, and more live launches planned for 2020.[26]

---

25. Emily A. Vogels, *Digital Divide Persists Even as Americans with Lower Incomes Make Gains in Tech Adoption*, Pew Rsch. Center, June 22, 2021, https://www.pewresearch.org/fact-tank/2019/05/07/digital-divide-persists-even-as-lower-income-americans-make-gains-in-tech-adoption/.

26. FCC, 2020 Broadband Deployment Report (2020) (citations omitted), https://docs.fcc.gov/public/attachments/FCC-20-50A1.pdf.

At the same time, the chair of the FCC in 2020 also acknowledged that while "97% of Americans have access to high-speed fixed service . . . [i]n rural areas, that number falls to 65%. And on Tribal lands, barely 60% have access. All told, nearly 30 million Americans cannot reap the benefits of the digital age."[27]

The government defines "access" as the physical availability of service, which means the presence of infrastructure to provide broadband. The federal government numbers therefore overstate actual access, as many people live in areas that have broadband infrastructure but cannot afford a subscription.

In addition, the FCC data does not speak to the quality of service. Not all broadband is the same. Satellite broadband is often the best available option in rural areas, many of which lack fiber optic and copper wire network infrastructure. When it is functioning, the quality and speed are high, but service can be impacted by cloud cover and other atmospheric disturbances, causing intermittent connection problems. DSL broadband, which runs over traditional phone lines, is widely available in the United States but the strength and reliability of a user's DSL connection depends on how close the user is to the switching station that serves as the node of the DSL service; quality and connectivity decline the farther the user is physically located form that node. These issues of connection quality are less important during jury selection, which is a relatively short portion of trial and requires relatively less combined upload and download connectivity by the juror, than they are during an online trial, when ongoing robust access is important to ensure that a juror is able to fully receive all of the evidence.

Unlike the FCC, the U.S. Census does collect data on households that have a subscription to broadband internet. The 2016 data shows that, on average, throughout the United States 81.4 percent of the households have a subscription to broadband internet service. There is significant regional variability, with the proportion of households with broadband subscriptions ranging from a high of 87.4 percent in Washington state to a low of 70.7 percent in Mississippi.[28] In general, rates of household broadband subscription are highest in the West and lowest in the Southeast.

27. FCC, *Chairman Pai Initiatives Archive*, https://www.fcc.gov/ajit-pai-initiatives-archive (last updated Jan. 20, 2021).

28. RYAN, *supra* note 23.

"Off-Line" Americans

Regardless of the availability of internet in their communities, there is still a portion of the American citizenry that does not use the internet, although this number has continuously declined since the advent of the internet in the late 1990s. A 2021 survey conducted by the Pew Research Center, for example, reports that nationally, 7 percent of American adults say that they never use the internet.

Research suggests that the primary reasons people do not use the internet are an inability to pay for broadband service and a discomfort or lack of skill in using the technology. According to the 2021 Pew report, age is the demographic variable most strongly associated with non-internet use, with about 25 percent of those over 65 saying that they never go online. Other factors related to non-internet use are educational attainment and income, with adults who live in households earning less than $30,000 a year or who have high school educations or less more likely to report no internet use.[29]

Summary

The data suggests that the large majority of Americans have access to adequate technology to participate in online jury selection or an online trial but that specific socioeconomic and racial groups may lack the needed technology. However, the same social characteristics associated with lack of computing devices and broadband access—advanced age, low educational attainment, low income, minority status—are also linked to a lack of participation in in-person jury service. In other words, those who are excluded are likely to also experience significant barriers to attending in-person jury service.

I would suggest, therefore, that the potential exclusion of people from these groups from participation in an online trial is not an argument against online trials—it is an argument for addressing the structural barriers that stand in the way of full citizen participation in jury duty and the justice system.

As we will see in the next section, the experience of judges and court administrators who have been in the trenches trying online cases is that solving the problems of access is actually much *easier* when conducting trials online—because it is easier to provide a computing device and

29. Andrew Perrin & Sara Atske, *7% of Americans Don't Use the Internet. Who Are They?*, Pew Rsch. Center, Apr. 2, 2021, https://www.pewresearch.org/fact-tank/2021/04/02/7-of-americans-dont-use-the-internet-who-are-they/.

internet access to jurors than it is to solve the more complex problems that stand in the way of in-person jury service.

Jury Pools in Online Trials

A handful of states and jurisdictions around the country have been on the cutting edge with experiments with online jury selections and online trials. These include some of the larger court systems in the country, including Hennepin County, Minnesota (home to the Minneapolis-St. Paul metropolitan area); King County, Washington (home to Seattle); Alameda County, California, and San Francisco, California (in the San Francisco metropolitan area); Maricopa County, Arizona (in the Phoenix metropolitan area); Florida; and Texas. This section draws on data and information from judges who have tried online jury trials in each of these venues, some through direct interviews by my colleague Marc King, PhD, and I and some through panel presentations and interviews conducted by other members of the Online Courtroom Project, supplemented with media accounts.[30]

The reports from these online courtrooms around the country are consistent: judges are seeing higher rates of jury service and more representative juries in online jury selections and online trials.

Some Judges Are Opting for Online Jury Selection and In-Person Trials

Jury selection is the portion of trial that poses the largest public health hazard, as it requires the largest assembly of people. During the year

30. With the assistance of Marc King, PhD, I conducted interviews of the following judges for this chapter: Judge Emily Miskel, 470th District Court, Collin County, Texas (Apr. 9, 2021); Judge Alan Albright, U.S. District Court—Western District of Texas (Waco) (Apr. 8, 2021); Judge Sean O'Donnell, King County Superior Court (Seattle) (Mar. 30, 2021). We also interviewed Texas State Court Administrator David Slayton (Apr. 1, 2021). In addition, I relied on comments made in the Online Courtroom Project's online summit by the following judges: Hon. Matthew W. Williams, King County Superior Court, Washington; Hon. Pamela Gates, Presiding Civil Department, Maricopa County, Arizona; Hon. Brad Seligman, Alameda County Superior Court, Alameda, California; Judge Jennifer D. Bailey, Circuit Civil, Dade County, Florida. Fellow Online Courtroom Project members Jeff Frederick and Karen Lisko have also generously shared additional information from their collaboration with Judge Gates to develop research-based best practices for technology-aided trials.

2020, when the epidemic was at its peak and vaccinations and treatments were not yet available, this led a number of judges to choose to hold jury selection online and then, once the jury was selected, to conduct the remainder of the trial in person, in a socially distanced courtroom.

This hybrid method is an interesting option now and in the future, as it brings the benefits of an online proceeding to the large majority of jurors while maintaining the traditional in-person trial procedures that so many of us in the justice system value. It also makes it possible to minimize the potential health risks of voir dire during criminal jury selection and satisfy the Confrontation Clause.

There is no question that online voir dire and jury selection is a different experience than in- person voir dire and jury selection. However, all of the judges who I have interviewed or heard speak on this issue agree that the benefits they have observed far outweigh the loss of in-person presence and any benefit of having all jurors hear the entire voir dire. Many of the benefits they mention go specifically to the issue of representativeness:

- Show rates are higher (which is evidence that it is generally easier for jurors to participate online rather than in person). Judges particularly note that more jurors who might have claimed hardships due to childcare or work issues with in-person service are appearing.
- Jurors do not have to take time and make arrangements to travel to and from court, which allows jurors without personal transportation to participate more easily.
- Jurors who are parents or caretakers of family members can participate from home and still provide care, or require much less outside assistance to provide care during jury selection periods.
- If judges schedule jurors to appear in panels of eight to 15 rather than requiring all jurors to attend the entire voir dire, jurors spend much less time in the jury selection process, making it easier for them to continue working and engaging in other life activities around the jury selection process.
- Jurors feel more comfortable discussing personal matters from home and appear to be more candid and complete in their voir dire responses.

Another significant benefit includes the higher quality of voir dire due to the use of supplemental questionnaires. (For a discussion of using supplemental questionnaires in online jury selection, see Chapter 15 in this volume, Online Jury Selection. See also http://t.ly/wskyN for a sample supplemental questionnaire for civil cases; and Appendix C for a list of recommended questions to ask jurors regarding technology and internet access.)

These benefits lead many to believe that, even post-COVID, some if not many courts will choose to hold jury selection only online, to then hold an in-person trial once the jury is sworn. Judge Miskel from Texas comments:

> My theory, going forward, is that we will see, when the public finds out it is possible for them to participate in jury duty remotely, the demand will come from these people who want to serve. The courthouse is built for my convenience, as a judge—not theirs. I love to have people come to my chambers and appear before me for me to solve their problems, but I am not here for my convenience. I'm here to serve the people of the state of Texas. I'm not some fan girl of virtual trials looking to sell them to everybody. I bend myself over backwards to do it the harder way [online] because it's working and it's better in some ways.[31]

In discussing best practices to ensure representative jury pools, it is also important to recognize that holding only jury selection online is an option because the technology and access challenges that may need to be overcome to hold a successful online jury *selection* and ensure a representative jury pool are more modest than those that may need to be overcome for a successful online *trial* that allows full participation for all selected jurors. Jurors need fewer data minutes, as jury selection is time-limited, especially if a judge schedules voir dire in panels; they also typically need less screen real estate, as during jury selection, it is more important for a juror to be seen than to see what others are doing or presenting. Jurors may therefore be able to fully participate successfully with only a smartphone rather than requiring provision of more robust computing and internet connection options. Even if the parties or court need to provide equipment and data access, it costs considerably less to provide that access for only a day of jury selection than for a week or more of trial.

Data from Online Trials

State and federal practices on collecting juror data vary widely by court and sometimes by judge, as noted above. Most courts do not collect

31. Interview with Judge Miskel, *supra* note 1.

demographic statistics to match the representativeness of their pools against census statistics. However, all courts collect official statistics on juror show rates in response to jury summons. Judges also pay attention to the mix of jurors who show up in their courts and many take informal notes on the composition of jury pools and seated juries and discuss their experiences with colleagues. This section discusses data found in published reports as well as anecdotal data from interviews with judges and court administrators. Because we have the most detailed information about Texas and King County, Washington, I present separate sections on these two jurisdictions first and then present additional anecdotal data from other jurisdictions.

Texas

The state of Texas has been the national leader in holding online court proceedings and held 45 full online trials in the past year. The state now averages two to three online trials per week.

David Slayton, the administrative director of the Texas Office of Court Administration, is the person in charge of the state's development of online trial protocols and he spoke to us at length about his office's experiences with online trials. He also shared with us the report that he and his office published in August 2020, *Jury Trials during the COVID-19 Pandemic: Observations and Recommendations*, which is one of the sources of the best practices I present later in this chapter and contains extensive data on the Texas experience with online trials. Slayton is also co-chair of the national Joint Technology Committee, which is a joint committee of the U.S. state and federal courts whose mission is to develop and promote technology standards for the courts.

While initially Slayton shared the concern that some jurors might be excluded from jury service due to lack of access to technology, his experience assuaged his concerns: "I don't think we're excluding more people than we were before. In fact, we know it's the opposite . . . what we're seeing is *increases* in rates of jury service. I think this is an underappreciated important thing that we're learning."[32]

Prior to the pandemic, Slayton says that nationally, the average show rate of summoned jurors was about 40 percent; the average show rate of summoned jurors in Texas was 25 percent. Those who showed up in

32. Interview by Marc King with David Slayton, Administrative Director, Texas Office of Court Administration (Apr. 1, 2021) [hereinafter Interview with David Slayton].

Texas were typically "older, affluent, white, so not completely representative of the community." With online jury trials,

> what we're seeing with virtual jury service is an exact replica of what we're seeing with other appearances, which is people are showing up and appearing. We are seeing courts that [prior to the pandemic] had 20, 30 percent appearance rates are having 70, 80 percent appearance rates. In our most recent Travis County case last week, only one juror who was summoned did not appear. The rest of them all appeared. And the jurors appear from the comfort of their home. They're relaxed.[33]

Slayton specifically noted that the number of Black and Hispanic citizens showing up for online jury duty is considerably higher than he has seen with in-person jury service. He recounted a story of a recent family law case, a termination of parental rights case, heard by a jury in Travis County:

> In Travis County, Austin, . . . there's a fairly significant African American population, but historically they've been way underrepresented in showing up for jury duty. We can talk about the causes of that all day long. There may be historical causes. There may be just, you know, barriers that get in the way, whatever it may be. The jury ended up being composed of four African Americans in a case where the state was seeking to terminate the parental rights of an African American family. In the past, there would not have been black Travis County residents on that jury. And just think about the procedural fairness perspective. They're looking at the jury on your screen and see people who look like them. I think from an accessibility standpoint, it's been huge.[34]

Judge Emily Miskel, who presides in Collin County, a suburb of the Dallas-Fort Worth area, has spoken with the district clerk in her county about the differences the clerk's office has observed between pre-COVID in-person jury pools and online jury pools. The clerk's office reports that pre-COVID, the median age of jurors was 48 and pools did not match the demographics of the county, while the median age of the online jury pool is 41 and the demographics match.

Judge Miskel also regularly speaks with colleagues about their experiences, and reports that her colleague Judge Roy Ferguson, who presides

---

33. *Id.*
34. *Id.*

over a vast area that covers five rural counties in West Texas, reports similar improvements in the representativeness of his jury pools—just the kind of counties where one might expect residents to have challenges with connectivity and computer access. His counties all have large Hispanic populations. Pre-COVID, he saw low response rates to jury summons, which he chalked up partly to having a large district and many people having transportation problems. Previously, his juries were frequently composed entirely of retired white people. Now, with online jury selection and trials, he is seeing many Hispanics showing up for service, eager to serve.[35]

With regards to access to technology, Slayton notes that the Texas Supreme Court

> has issued its order, a requirement that if we're going to do virtual trials, we have to provide access to technology to those who don't have it. So we actually ask that as part of the prequalification process. Do you have access to technology? Do you own a webcam? Do you have unlimited Internet access, things like that. And if they don't, we literally deliver to their homes an iPad with cell service.[36]

King County, Washington (Seattle Metro Area)

The COVID-19 virus was first officially acknowledged to be present in the United States in Washington state. Washington state has also been at the forefront of experimenting with online trials as a way to keep its justice system operating during the pandemic, in both its state and federal court systems. King County, Washington—home to Seattle, Washington's largest city as well as home to many cutting-edge tech companies, including Microsoft, Boeing, Amazon.com, Cray, and Redfin—has been one of the counties pushing hardest at the envelope in the state court system. Since late July 2020, King County has conducted hundreds of jury trials and thousands of bench trials. About 100 of these have been full online jury trials using Zoom. Some have been hybrid trials with online jury selection. These jury trials, whether online or hybrid, could be held under conditions of COVID only because the courts conducted jury selection online, with prospective jurors appearing from home.[37]

35. Interview with Judge Miskel, *supra* note 1.
36. Interview with David Slayton, *supra* note 32.
37. Jim Rogers, *Justice Delayed: COVID-19's Staggering Criminal-Case Backlog*, Seattle Times, Apr. 7, 2021, https://www.seattletimes.com/opinion/justice-delayed-covid-19s-staggering-criminal-case-backlog/.

To dig more deeply into the details of the King County experience, we conducted interviews with one of the judges who had been holding Zoom trials, Judge Sean O'Donnell, and his bailiff, Ann Brockenbrough, about their experiences. Ms. Brockenbrough also kindly collected and shared data with us from across eight King County courtrooms on their experiences with online jury selection and online trials during 2020.

As of our interview in March 2021, Judge O'Donnell estimates that he has conducted nine trials using online technology during the pandemic—four complete online trials, from beginning to end; and five "hybrid" trials, where the court used Zoom to conduct jury selection online and then held the remainder of the trial in person. It took an average of 1.5 days to complete jury selection, but only about one day of that was spent in actual selection activities.

King County has set up workstations in the courthouse law library for jurors who lack the technology to participate in online trials. In all of his Zoom trials, Judge O'Donnell himself never had a juror request that option, although he has colleagues whose jurors have. He notes:

> Maybe 1 out of 14 of my jurors are on their cell phone. Usually, they are on an iPad, desktop or laptop. It's rare that I have someone using a phone. . . . It would be great if I had a stack of iPads I could send out as loaners and we could program it so that it had 6 hours of free data and the iPad could be programmed to access only the Court-access website. But a practical solution if a juror wants to serve but has technical challenges is for that juror just to choose to come and serve in person. During COVID, we have been able to maintain social distancing without [a] problem because there have been so many fewer jurors and other people in the courthouse.[38]

Judge O'Donnell's general conclusion about conducting a trial online is that

> it's different. Not worse, but different. The cases I've handled, jurors are being incredibly careful, diligent and attentive to their jobs as jurors. I will say that is a common feature with in person juries. They are careful, attentive, thoughtful, collaborative. All of those descriptors, I've seen in online juries. Whether they

38. Interview by Sarah Murray and Marc King with Judge Sean O'Donnell, King County Superior Court, Washington (Mar. 30, 2021).

are going to go have a drink together after the trial is over—I couldn't tell you.[39]

He also said based on his experience as well as his conversations with colleagues, "My impression is that the immense convenience of online trials has increased the number of people who serve. To those who say, 'You're going to lose jurors!'—persuade me that they were going to come anyway."[40]

Ms. Brockenbrough collects data for Judge O'Donnell and also collected data from seven other King County courtrooms on their experiences conducting jury selection by Zoom in 2020. Jury pools varied in size from 80 to 200, with the typical pool composed of 120 jurors. Questionnaires were sent to jurors using Microsoft Forms. The average response rate to the questionnaires, across approximately 30 online trials, was about 82 percent. When an introductory email was sent first, response rates went up to an average of 91 percent; when a reminder email was sent, response rates went up to 87 percent.

On average, only 2.2 percent of jurors were unable to participate in jury selection due to technology issues. Judges who gave jurors the option of attending in person if they chose experienced no loss of jurors.

Some judges reported jurors' lack of a dedicated, quiet space in the home was a more significant problem than lack of access to technology and connectivity. This echoes data gathered by Judge Gates in Arizona, discussed further below, which shows that non-technology reasons are greater barriers to participation in online trials than lack of technology.

Judge O'Donnell reports that he has seen more diverse jury pools in his courtroom and hears that his colleagues are experiencing the same. He expects the county to continue to offer online jury selection as an option in both civil and criminal cases even after public health measures no longer require social distancing. Ms. Brockenbrough comments that, based on her experiences, in King County: "Technology doesn't appear to be a huge barrier to participation for the vast majority of people; bailiff technical support of jurors is helpful here, and so is the option for people to come to the courthouse to participate if they wish."[41]

39. *Id.*

40. *Id.*

41. Email from Ann Brockenbrough, Bailiff, King County Superior Court, to Marc King and Sarah Murray (Mar. 30, 2021, 6:07 p.m.).

Maricopa County, Arizona

Judge Pamela Gates in Maricopa County has been working with Karen Lisko, PhD, and Jeff Frederick, PhD, of the Online Courtroom Project to study best practices for conducting online trials in her county, which is the fifth largest judicial district in the country. Even before COVID-19, Maricopa County used an online e-juror portal to send jury summons and administer pretrial questionnaires online to jurors being summoned for in-person jury service. Judge Gates reports a response rate of nearly 80 percent to these online summons. In May 2020, Judge Gates worked with Drs. Lisko and Frederick to develop a more robust online questionnaire that included new questions to gather information about prospective jurors' ability to serve in an online proceeding.

They surveyed approximately 40,000 Arizona jurors: 62 percent report that they meet the conditions the county sets for jury service from home, which includes:

- Access to a quiet space that is free from interruption
- Reliable internet service
- Access to an appropriate electronic device
- The physical ability to sit through a trial

The most commonly cited reason for an inability to serve, cited by 31 percent, is that they would have interruptions (63 percent of those who say they would have interruptions are female; 37 percent are male).

- 22 percent of those surveyed said they would have physical challenges with watching and/or listening, and most of those were 70 years or older
- 18 percent said they would lack quiet or private space suitable for jury service
- Only 11 percent said that they lacked reliable internet access
- Only 10 percent of jurors lacked a device

In the Online Courtroom Project's summit on online trials, Judge Gates reported that Latino jurors experienced more difficulties than other ethnic groups participating in an online trial.

Florida

Judge Jennifer Bailey of the Eleventh Judicial Circuit in Florida had jurors appear in panels of 15. She reports higher show rates in completing the court juror questionnaire than pre-COVID and noted that almost 100 percent of those who filled out the online juror questionnaire showed up for the online jury selection. She also commented that she felt the level of attention during voir dire was more intense than what is typical during in-person voir dire in court.

Best Practices to Ensure Representative Jury Pools in Online Proceedings

The best practices described here address only how to ensure that potential jurors are not excluded from service based on a lack of access to computers or broadband internet service and do not address any of the many other barriers that prevent some Americans from serving on juries. They also do not address best practices for voir dire and online jury selection—those are addressed in Chapter 15 in this volume.

- As part of the jury summons, explain that the jury selection or entire trial will be held online and discuss the arrangements that the court is making to ensure that every juror called will have access to appropriate technology—or give jurors without access to technology the option of appearing in court for in-person voir dire.
- As part of the summons, explain the preparations that the juror should make to ensure privacy and full focus on the voir dire process:
  - Find a quiet space, ideally one with a door that can be closed.
  - Inform household members that you will be unavailable for the period of time scheduled for voir dire except for emergencies. If necessary, arrange for childcare, pet care, or in-home health care for the period of time you are scheduled to participate.
  - Turn off all other electronic devices, including televisions and radios.
  - Close out of all other programs besides the program required to connect to the court.
  - Turn off any noisy fans, air filters, or other devices that might interfere with your audio transmission.
- Administer a short background questionnaire to all jurors to determine their access to technology at least two weeks before jury selection begins. A sample questionnaire is included at the link given in Appendix F and directly at http://t.ly/wskyN.
  - Questionnaires should include questions about a juror's types of computer equipment and internet access.
  - For ease of administration and data management, host the questionnaire online.

- Format the questionnaire so that it can be taken on any type of computing device, including a smartphone.
- In the jury summons, let jurors know of local library locations where they can access the internet to take the questionnaire online if they do not currently have access to a device or the internet.
- Also make the questionnaire available in a paper copy by request, by mail or at the courthouse for in-person pickup.
- Send reminders to jurors to complete the questionnaire.

- If a juror has smartphone access only, or has limited high-speed internet access, consider options to give that juror access to more robust technology as your structure might need, which might include:
    - Sending a "loaner" tablet or laptop to the juror
    - Sending the juror a card for additional data minutes
    - Sending the juror a "loaner" hot spot that allows for high-speed internet access
    - Making arrangements for the juror to come to the courthouse or government office that has internet-connected computer workstations during the course of jury selection
    - Making arrangements for panels of jurors without appropriate access to come to court for in-person jury selection (socially distanced as necessary)
    - If the remainder of the trial is to be held online, once the jury is empaneled, making arrangements to provide the juror with appropriate computer technology and internet access or allowing the juror or jurors to attend the trial in person
- Schedule jurors to appear in panels of eight to 15 rather than requiring the full pool to sign in and be online together for the entire voir dire.
- Before jury selection is due to start, have the clerk or a trial tech conduct tech tests with panels of eight to 15 jurors to ensure that jurors' connections and equipment are adequate and that jurors understand the process.

If jury selection is being held online, as discussed earlier, a smartphone may be sufficient for a juror to engage in the voir dire process, depending on how the process is managed, as during jury selection the juror's ability to see others is less important than the ability of the court and the lawyers to see the potential juror.

Once a jury is empaneled, check in again with jurors about their access to ensure that they have sufficient equipment and broadband access at home to continue with the trial, or if they will need alternate access through a library or other public facilities the court has arranged. If a juror has experienced connection or equipment challenges during voir dire, it is likely the juror will need new equipment and/or additional broadband access to successfully participate in the trial.

Conclusion

Although access to computers and broadband internet access has continually increased since the introduction of both technologies to the United States, there remains uneven distribution of access to both. Various data sources agree that lower-income Americans, Native Americans who live on tribal lands, rural Americans, and older Americans in particular have less access to technology and may also have lower comfort levels with using technology. There are also regional differences in levels of access to technology. However, lack of access to technology is not the only issue that impacts citizen ability to participate in jury duty and jury selection. Data shows that the same portions of society who face technology access challenges face challenges in showing up for in-person jury trials as well, due to such factors as lack of access to transportation, employment in low-wage jobs that do not pay for jury service, and alienation from a justice system that they have experienced as unjust or unrelated to their own experience. These challenges of representation are therefore not unique to online trials but reflect systemic inequities and problems in our society. The experience of courts around the country who have experimented with online trials in this past year suggests that solving the technical challenges to ensure equal access and more representative juries outlined in this chapter is likely to prove easier than solving the broader systemic challenges that have long presented barriers to full citizen participation in in-person jury service.

Judges and courts have met the challenges of ensuring equal access to all potential jurors for online trials in a variety of ways, including using existing government and court computing resources and, in civil cases, requiring parties to provide needed technology to jurors. Judges report that show rates for online jury selection are higher than for in person jury selection and jurors are enthusiastic about the increased ease of participating in this vital aspect of American citizenship.

Like Judge Miskel, I am not a "fan girl" of online trials. I am a fan of justice. Online trials present both challenges and opportunities; we also still have work to do to recreate the rituals of justice in an online setting in a way that creates the gravitas and social commitment to the trial process that has been developed over many years of refining the rituals of in-person trials. However, with regard to the issue of representativeness, based on the experiences of the past year, the evidence is clear. Online jury selection, whether as the only online component of a trial or part of a full online trial, shows the potential to *increase* the representativeness of American jury pools and the participation of Americans in jury service, and to help revitalize what has become an increasingly moribund trial jury system.

Part Two

Best Practices in the Online Litigation Process

Chapter 9

Online Client and Witness Meetings

Ken Broda-Bahm

The idea of using web-conferencing for a meeting, of course, was not new in 2020. At the same time, many of us did not have the habit of using it very regularly. For routine contacts, we would use email or the telephone, and for important meetings, we would come together in person. That latter habit changed very quickly in the early spring of 2020. Suddenly, people who had made occasional use of the Skype or FaceTime application on their laptops or phone, were now regularly meeting with groups of individuals within the now-familiar grid of a web-conferencing application like Zoom, Webex, or Microsoft Teams.

It is safe to say that, through much of 2020, 2021, and 2022, even as hearings and trial dates may have been delayed, the meetings continued. Client and witness meetings have probably been among the most common virtual activities embraced by litigators and others working in the trial process. As a result, many of us have come to the realization that it is actually pretty easy to create an appointment, send a link, and have the meeting online. The benefits of face-to-face interactions, most of them at least, have ended up being preserved. An added bonus is that we are saving the time that was previously devoted to driving or flying to the meeting, along with the expense of travel. While many are likely to return to a strong preference for in-person meetings post-pandemic, the ease and the economy of the online setting for client and witness meetings suggests that this innovation will probably persist for at least some of those meetings.

In this chapter, I will share current thoughts on the best practices for both online legal meetings generally, and for meetings with the goal of witness preparation in particular.

Online Meetings Generally

Regular meetings are likely to be among the least problematic applications of web-conferencing technology to the legal process. In general, we should conduct them the same way we conduct any other virtual meeting. At the same time, we should give some consideration to the communication differences in a video-mediated environment (see, in particular, the recommendations in Chapter 12 on communicating in a virtual forum).

The following are recommended best practices for conducting general client meetings online as part of the litigation process.

Provide a Reminder and a Record on Confidentiality and Attorney-Client Privilege

In an in-person meeting, you know who is present, but in an online setting, you will need to confirm:

- Is it correct that I am talking with the three of you, and there is no one else in the room, and no one else within earshot?
- Is it also correct that no one is making a recording of this conversation or otherwise making a record of this meeting?

Do an Equipment Check in Advance

Asking a member of your team to conduct a quick online connectivity test before the meeting can save everyone some time and stress. Even taking just ten minutes to ensure that each meeting participant has what is needed—computer, speaker, microphone, camera, sufficient internet speed, and access to the platform—will help, particularly for those who may not be used to daily web-conferences. Ensure that each participant can be easily seen and heard and understands the basics (such as muting, unmuting, and switching the camera on and off). This is also a good time to encourage participants to use a computer desktop or laptop, if available, rather than a smartphone or tablet.

Make Meetings Shorter, but More Numerous

It has been noted that there is a "Zoom fatigue"[1] that can set in after a couple of hours. Because we pay attention differently during a web-conference, focusing on the meeting can be uniquely taxing when it is conducted online. While an in-person meeting associated with litigation

1. Jeremy N. Bailenson, *Nonverbal Overload: A Theoretical Argument for the Causes of Zoom Fatigue*, 2(1) TECH. MIND & BEHAV. (2021).

planning could run for many hours, or even all day, it is pushing the limits of our attention and technological comfort for an online meeting to last that long. Instead of pushing through to complete the work in one sitting, it is a better practice to hold shorter sessions, but more of them if needed. The ease in scheduling and starting an online meeting, and the fact that no one needs to travel, makes it easier to say, "Let's call it a day for now and find an hour next week to continue."

Potentially Include More People

The nature of the online meeting space also makes it easy to include more people when it makes sense. Clients, co-counsel, paralegals, insurance representatives, and others may want to listen in just in case relevant issues come up. They can also do so without "crowding" the space: if anyone's role is simply to assess the status, or to be a just-in-case resource, then they can keep their camera off in order to keep things simpler and more direct for the attorney and the client.

Make Sure All Active Participants Are on Camera

In some online meeting scenarios that we have seen, some of the people are meeting in a conference room in person, while others are logging in via a web-conference. One disadvantage this has is difficulty in setting up a good camera position: you either leave some speakers off-camera in order to focus on others, or you end up having a side view of everyone. You could also have the problem seen during the pandemic: if you are following the health advice, the in-person participants will need to be masked. My experience is that the better online meetings are fully online, where every person who will be speaking has their own laptop and camera.

Best Practices Specifically for Witness Preparation Meetings

All of the best practices for online meetings generally apply to witness preparation meetings as well, but the act of preparing a witness for trial or deposition testimony is a unique activity that also deserves special attention in an online environment. Whether the testimony itself is to be delivered in person or virtually, an online meeting to prepare that testimony needs to be handled carefully. The following are my recommended best practices for the specific use of online communication to prepare the witness.

Address Confidentiality and Protected Status

When the witness is a fact witness or a client, then attorney work product or client protection is likely to apply, and witnesses should be apprised of that condition and should understand that, even though their words are traveling over an internet connection, they still enjoy the same protection that they would if the meeting were being held in a private office.

Set the Foundation

Emphasize to your witness that all of the basic considerations of effective communication still apply in this online setting, and some deserve special emphasis. For example, witnesses should get used to sitting up straight and looking at the camera and maintaining a calm demeanor along with an alert and focused posture and facial expression, while minimizing unnecessary movements (rocking or fidgeting), which can be exaggerated by the camera.

As part of the foundation, Holland and Knight partner Daniel Small, the author of the American Bar Association's guide to witness preparation,[2] recommends sending the remote witness three items before your session: a "rules booklet" (including your preferred "dos" and "don'ts"), an "issues outline" focusing on the witness's areas of testimony, and an "exhibit binder" including any documents the witness might be shown. Sending these to the witness in advance, electronically or in hard copy as the witness prefers, helps to ensure that the witness has everything needed before you start.

Prepare the Witness for Questions about Off-Camera Resources

Confirm that your witness is alone in a room, with no one else able to observe or overhear. They should also be unaided by any notes or other off-camera resources, other than the documents you are asking them to review.

For witnesses who will also be testifying online, prepare them to answer questions about their environment:

- Do you have any notes in front of you?
- Do you have any notes anywhere in this room?
- Are you alone in this room?
- Is anyone other than the four of us on this web-conference able to hear your responses?

2. Daniel I. Small, Preparing Witnesses: A Practical Guide for Lawyers and Their Clients (2004).

- Do you have any windows or programs open on your laptop other than Zoom?
- Do you have the ability to rotate your camera 360 degrees so I can just see the space that you are in?
- If anything changes during this testimony and information does become available from some source in your environment, will you let me know?

Pay Attention to Setup and Camera

When the testimony that you are preparing for is online, or may be online, then it makes sense to spend some time talking about the setup: a neutral background, plenty of light in front of the witness (not behind and backlighting the witness), the camera at eye level, and the upper body centered and filling most of the frame. When the witness is likely to be self-conscious or distracted, you might consider having the witness turn off the "self-view" on their own monitor. Even if the testimony is likely to be in person, it still makes sense to pay attention to the optics for the preparation meeting, both because some of those factors (like posture) will apply in an in-person setting as well, and also because you simply want to practice good communication in a meeting about good communication (see also Chapter 12 on communicating in a virtual forum).

Check Their Space in Advance

Check in with the witness, ideally before your meeting, to ensure that the witness has an appropriate location for the preparation session, and for the ultimate testimony as well, if that testimony is online. Encourage your witness to find a space with a table and a neutral background, such as a blank wall. Witnesses should not typically apply a virtual background because, when not professionally used, they can distort the witness's video image. If the witness will be testifying online, you can also use the preparation session as a way to assess and to revise the space that will ultimately be used.

If Preparing for Remote Testimony, Use the Same Platform

Depending on your witness's level of familiarity with the technology, there may or may not be a learning curve for them in participating in the preparation session. At a basic level, they should know how to log in, to adjust the video and the sound, to mute and unmute themselves, and to change their view of other participants (i.e., "gallery" or "speaker" view). These tasks are generally simple but can still vary from platform to platform. Thus, there are good reasons to prepare using the same platform that will be used for testimony.

Play to the Uniqueness of the Format

One reality of online communication is that glitches do occur: someone's video might freeze, or their voice might cut out for a moment. At times, the problems are large and obvious, and might require your witness to know to log out and log back in, or even reboot their technology. At other times, the problem might be so small as to be almost unnoticeable—a small hiccup, but large enough to erase a key word from a question. But in a way, these limits can create an advantage. The benefit is that the technological uncertainties encourage online witnesses to do what every witness should do:

- Ask for clarification whenever there is any uncertainty over what is being asked. Slow down and take your time in answering. As Dan Small notes, "Everything needs to go slower on video. Everyone involved, including the witness and counsel, need time to think, react or object."[3]

Keep Some Informal Conversation

There is something about the online-meeting medium that conveys a "Let's get down to business" attitude. This is perhaps the reason that online company meetings have generally ended up being a little shorter than their in-person counterparts. In a witness preparation session, however, there is an important purpose served by informal conversation and chatting: it allows you to assess the witness's baseline communication habits, and it allows the potentially nervous witness to get more comfortable with you and the process. So, leave some time for less formal communication.

Practice the Testimony

Use part of the online meeting to conduct a dry run of the expected testimony. Ask the questions you expect opposing counsel to ask, and let the witness try out an acceptable answer. Provide feedback on both the substance and the style of the testimony. The goal of practice is not to "script" the witness, but to instead simply build familiarity with both the

3. Dan Small, *The Remote Witness: Preparation Challenges*, LAW.COM DAILY BUS. REV., June 3, 2020, https://www.law.com/dailybusinessreview/2020/06/03/the-remote-witness-preparation-challenges/.

process and the areas where they are likely to be questioned. The dry run will also give the witness experience with the medium, for example, highlighting the need to pause before answering and to not overlap the questioner.

Do Not Overload

In any witness preparation meeting, a common pitfall is trying to do too much in one meeting. If you tell a witness literally everything that they could be doing better, you are probably overloading them. For example, meetings that mix a focus on the substance (What is our position?) as well as execution (How do we communicate it?) may be trying to do too much. If we treat the online preparation session as a single precious opportunity to get it all done, then we may be overreaching, especially when using technology that makes it a little harder to maintain attention. The medium makes it easy to create separate meetings: discuss our stance in one meeting, then practice the testimony in another. Later, consider a full mock deposition.

Consider Recording Practice

Finally, there is often value in having witnesses watch themselves when practicing testimony, since they can appreciate some of your advice more effectively when they are able to see both the good and the bad examples for themselves. To do that online, you need a way to effectively record and play back. Naturally, you do not want to record the entire session, but you do want to record examples of sample testimony so you can offer an example and a critique. With most platforms, you can use the "record" feature of your web-conference software platform, selectively turning it on or off, and then sharing your own screen as you play back. In practice, you might find that playing back video overwhelms your software or your bandwidth, so you might also have an alternate way to share the video (by uploading it to a secure Dropbox site, for example, or even simply playing it on a separate device, like an iPad pointed at your camera). Depending on the given attorney's level of confidence in client and work product protection, of course, some may wish to avoid recording and simply rely on verbal feedback. In that case, one advantage of the medium is that, even without a record, the witness still has the chance to see themselves on the screen.

Conclusion

To be sure, there are some sacrifices in a virtual environment. Given the sensitivity of client meetings as well as the interpersonal demands of preparing a witness for testimony, many will want the most important of those meetings to be in person wherever possible. Still, despite the reality that some communication nuances are likely to be diminished in an online setting, it is still the case that much of the work that is done with clients and witnesses can be done to a relatively high standard in an online environment. Following the practices recommended in this chapter, attorneys conducting online meetings and preparation sessions can expect to gain most of the value of an in-person preparation session.

Chapter 10

Taking and Defending Online Depositions

Geoffrey Vance and Robert Kopka

Introduction

Prior to the pandemic, remote depositions were rare, usually reserved for busy doctors or far-away witnesses. COVID-19 changed that. Likely forever.

Since the beginning of the pandemic, attorneys have been forced to take and defend depositions remotely. The technology necessary to take remote depositions is an ever-evolving challenge. More importantly, attorneys must learn to develop new interrogation skills to make up for lost opportunities to "size up" the deponent in person. Attorneys defending depositions must develop and employ new strategies to protect their deponents and enhance their appearance and credibility. In preparing this chapter, the authors interviewed dozens of busy civil trial attorneys at our respective law firms and gleaned dozens of helpful tips and recommendations that we proudly pass along to you. Our goal is to provide practical tips gleaned from the experiences of attorneys who collectively have taken hundreds of remote depositions. Their observations can be separated into two major categories: (1) adoption and mastery of the remote technology; and (2) adaptation to and implementation of new strategies arising from this once-unique format.

Adoption and Mastery of Remote Technology

A consistent comment from top civil litigators is that preparation and familiarity with the remote deposition technology is vital. Witnesses and lawyers need the right equipment. Lawyers need to be skilled in navigating both the hardware and the software necessary to conduct a remote deposition. There is no area in which familiarity and aptitude in navigating the technology is more important than in handling and referring to exhibits.

Equipment

Depositions should be conducted with fully functional laptops or personal computers, high-definition cameras and unobstructed sound devices, microphones, and headsets. Wireless sound devices should have noise-canceling functionality and should have enough battery life to survive the entire deposition. Attorneys should test all equipment well in advance and have available backup equipment in a convenient place, ready at a moment's notice. Court reporting services may have offices equipped with equipment for partially remote depositions, but when relying on others' equipment, similar precautions are warranted.

Deponents who do not possess proper computer equipment or adequate internet to participate in a remote deposition are a particular problem. Some national court reporting services provide laptop computers and portable Wi-Fi routers to ill-equipped deponents for a fee. They drop-ship the laptop and router directly to the deponent with clear instructions, with plenty of advance time for the defending attorney to test the equipment and prepare the deponent.

Exhibits

Handling and showing exhibits pose the biggest technological challenge during depositions. Lawyers who are not adept at showing and introducing exhibits electronically cause undue delay and inhibit their own ability to effectively conduct the deposition. Neither the witnesses nor the court reporter should be expected to operate the technology. It is imperative for the attorney taking the deposition to assure, in advance, that all arrangements have been made for the dispersion of technology to all relevant parties including counsel and witnesses. Moreover, it is not enough to assure that the deponent has proper equipment; one must also assure that all parties have access to the internet and a computer. Attorneys who rely upon court reporters to gather and show exhibits delay the deposition and waste a key opportunity to take control of the deposition.

Attorneys should know how to use the screen-share function and be ready to show witness exhibits using the technology. An attorney who can properly operate the technology will be able to perform functions like showing witness exhibits, zooming into certain aspects of an exhibit and even providing the witness with a highlighted portion of an exhibit with an ability to mark it up. An attorney who is facile and adept at operating technology is likely to intimidate the deponent and impress the opposing counsel and anyone who subsequently views the video recording of the deposition. Properly prepared exhibits can highlight themes and direct testimony toward favorable admissions. Conversely, an attorney who fumbles with the exhibits and the related technology loses the battle of perception with the deponent and counsel and squanders the opportunity to adduce evidence advantageous to his or her client.

Few, if any, jurisdictions require an exchange of specific documents to be used in a deposition. Certainly, only documents that have been disclosed to the other side may be used in a deposition, but counsel are free to choose exhibits randomly from the entire universe of produced documents and (ethically) "surprise" a witness during a deposition by reference to a document that the deponent was unprepared to confront. Remote depositions typically operate under the local rules of civil procedure and while disclosure of exhibits to be used at a deposition is rarely mandated by rule, it is increasingly advisable in a remote deposition context.

Prior preparation, consultation, and cooperation with opposing counsel, especially with regard to exhibits, will allow the deposition to proceed much more smoothly. If witnesses have an opportunity to review exhibits before the deposition, it is easier to draw their attention to certain exhibits on the screen. It is no longer practical for a witness to remotely review a 300-page document in order to stipulate to its authenticity or even before answering questions about a page or two of the document. Therefore, it is incumbent upon the attorneys to seek stipulations regarding the authenticity of exhibits. This is one of several instances in which the attorney willingly concedes the perceived tactical advantage of "springing" a document on an unsuspecting deponent in exchange for the orderly and efficient review of exhibits during the deposition. By exchanging and agreeing to the deposition exhibits in advance, the parties will facilitate a smooth deposition and a useful deposition transcript. Of course, even if you do exchange exhibits in advance, there may be times when you decide to show the deponent an exhibit you did not provide opposing counsel in advance of the deposition, especially when the deponent testifies to something you could not have anticipated when preparing for the deposition. You may also provide exhibits in a sealed envelope for the witness to open during the deposition. And with more effect if you are adept at the technology.

Video Recordings

Consult local rules to determine if, at all, video recordings of remote depositions may be used at trial. It is wise to seek a stipulation with opposing counsel to use the deposition video in court, for impeachment or other permissible purposes, if there is even a small possibility the case will get tried. It is unclear that a record of a remote deposition will meet the evidentiary standards of a video deposition, though we think it is likely most courts will conclude as much. Regardless, a stipulation would benefit the party noticing the deposition to ensure his or her ability to use the video deposition at trial.

Background and Lighting

It is important to note the enhanced importance of the background and lighting in every deposition. Where appropriate (especially when it benefits your client), provide witnesses with tripods, webcams, earbuds, laptops, LED ring lights, and even vinyl portrait backdrops to enhance your witness's appearance. The right equipment is key to being effective for both witnesses and attorneys. Reliable, high-speed bandwidth internet service is essential. A noise-free environment is crucial. A professional presentation provides your witnesses with the best presentation. A vinyl background can be important and useful to provide a neutral setting when a deposition may be taken in a home or business. The virtual backgrounds provided by Zoom and similar products are unsteady and not only make witnesses look cartoonish but diminish the processing power of the computer and significantly affect the quality of transmission. Recently, one client participated in a deposition in his garage, which made the vinyl backdrop an important part of the deposition.

Technical Glitches

Technical glitches can and should be expected to occur. Always have backups. Make sure that all attorneys in the room exchange cell phone information and can communicate in the event of an internet breakdown. Have a plan for various contingencies, with contingent equipment available if possible.

Technology is constantly evolving and so should you. We anticipate that the reduced cost and increased convenience of remote depositions will propel them to the mainstream, even after the pandemic subsides. Accordingly, the technology will also evolve. Those attorneys who continue to learn and master the latest technology will have a long-lasting advantage in taking effective depositions into the foreseeable future.

Adaptation to New Strategies

The attorneys interviewed for this chapter provided incredible insight into strategies they have utilized to take advantage of the changes necessitated by taking and defending depositions remotely. The attorneys provided recommendations for both interrogation tactics and defense tactics. In this section, we recount strategies you should consider employing in your own litigation practice as you perfect the art of taking and defending remote depositions.

From the defense perspective, preparation and practice is key. Credibility in a deposition is always tied to a dynamic appearance. Remote depositions tend to flatten out the deponent's profile, hide the deponent's body, and change the overall impression of the witness. A witness not properly trained or prepared can look suspicious and disinterested. Moreover, as remote depositions bleed into the afternoon and evening, video fatigue may set in. A witness may need breaks more often to recharge and be reminded of the importance of looking at the camera and speaking directly.

Witness Preparation

Preparation of witnesses includes rehearsal with the technology and a dialogue about contingencies in the event of a technology breakdown. Lawyers and clients must continue to develop videoconferencing experience. Smart practitioners will meet virtually with their witnesses at least several days, if not weeks, before the deposition and then several times thereafter to provide adequate recommendations, coaching preparation, and training. First and foremost, witnesses should be instructed to look directly at the camera as if the camera were asking the questions. It is natural for a witness to look at the questioner who is often on the screen disconnected from the camera, unless it is on the laptop. Even on a laptop, a witness can be directed to look directly at the camera. Looking at the camera enhances the credibility of the witness.

Practice with the witness. If deposition exhibits are expected, review those exhibits with the witness remotely, even going so far as to demonstrate the ability to mark up the exhibits. If necessary or appropriate, provide the witness with printed copies of the exhibits as well so he or she may spend the time necessary to read the full exhibit before answering questions about a part of it. This is another reason why agreeing to the exhibits well in advance with opposing counsel is a benefit to both sides. Stress the importance of pauses to allow for objections or just to digest and think of the answer to the question before speaking.

A number of litigators have observed that witnesses can often be more effective by video, if properly trained, looking directly at the camera, pausing before answering and answering only the question asked, with a sincere voice. These are all techniques that enhance credibility on video. Proper lighting is critical. A witness whose face is properly lit looks friendlier and more credible. A dimly lit face with a shadowy background connotes distrust and disbelief. Proper preparation will allow the witness to be more at ease testifying. Ensuring a basic backdrop will focus attention to the witness and avoid distracting attention to items in the room.

Defending Remote Depositions

Many practitioners believe they must be in the room to most effectively defend a deposition. This provides psychological and emotional support for the deponent and reassurance that the attorney representing them is right next to them, literally feet away. When the attorney needs instant access or the ability to communicate directly with the deponent, being in the same physical room provides a substantial advantage. When it is not possible to be in the same room as the deponent, some practitioners have devised a word or a motion of the hand as a signal that they may want to take a break or rethink what they said, though we can envision this backfiring, especially if it becomes obvious on the video. A better approach is probably to ask the witness, "Do you need a break?" and to make sure the witness knows in advance that when their lawyer asks them that, the answer should almost always be "Yes."

Regarding objections, making them in a timely fashion can be much more difficult in a remote deposition setting. It is best to seek stipulations with opposing counsel regarding objections before the remote deposition begins. Standing objections can be used to facilitate an easier transition to remote depositions. Fewer objections facilitate more effective testimony and a better deposition transcript. In other words, unlike live depositions, where time or internet delays are not an issue, it's much more difficult to rattle a witness with an objection. The delays inherent in remote depositions steal that thunder.

Taking Remote Depositions

Practical tips in taking remote interrogation of witnesses sometimes involves changing the way a practicing attorney interrogates the witness. Many attorneys lament the fact that they are no longer in the same room as the deponent, able to "look them over" and build a rapport in person. They can no longer watch them walk into the office or come into the building. They are no longer present in the room with them to observe their body language. One tactic is to ask the deponent to adjust the camera to the widest angle so that the entire room is in the picture or to rotate their camera to show the deponent's surroundings. There are many things that can be gleaned from looking at the room in which the deponent is giving the deposition, including additional phones, computers, and devices as well as stacks of paper or even other people. It is not improper to ask the deponent to show you the room in which he or she is giving the deposition. It is critical to make sure that the deponent is alone in the room and that other witnesses in the case are not in the room. Clearly the deponent should be asked with whom the deponent has spoken regarding the subject matter of the deposition.

Details in the case can sometimes become much more important in a remote deposition environment. While live depositions often allow the practicing attorney to get a better "feel" for the credibility of the witness, the credibility of the witness can be tested remotely by reviewing the testimony in detail, seeking inconsistency in documents or with the testimony of other witnesses. The credibility of each witness becomes much more important in remote depositions. Some lawyers and judges like videos better because the face of the witness is framed, making every smirk, eye roll, and other facial expression visible to the observer.

Attorneys report two opposite tactics to obtain key admissions when interrogating a deponent. One method is to disarm the deponent with charm. Recognizing that the deponent may be alone and unsure of the technology or the process, the interrogating attorney may take advantage of the distance between the deponent and his or her counsel. By appearing sympathetic and moving slowly at first, the attorney may gain the deponent's trust, leading the witness to make key admissions he or she may not have made if the attorneys from both sides were in the same room. Witnesses can be made to be more relaxed in a remote deposition and admissions can sometimes be elicited by giving the deponent a false sense of informality and security. If the deponent is not in the same room as the defense attorney, the practicing attorney taking the deposition can create a virtual rapport with the deponent and lull the deponent into a series of admissions.

Another tactic is to be aggressive and to force witnesses into defensive behavior. Defensive witnesses often appear non-credible and less than genuine. Witnesses who feel under attack may get angry and blurt key admissions, especially when separated from their counsel. Or, they may be lulled into admissions. A good tactic here is to prepare probing questions seeking a lot of detailed answers and then challenging witnesses on each specific detail to put them on the defense.

Other Considerations

One of the things lost in remote depositions is the ability of counsel to talk among themselves after a deposition. These pre- and post-deposition moments were often an opportunity to seek to narrow the issues in the case or move the matter toward eventual resolution. This was an important part of moving cases toward successful mediation or trial.

Litigating remotely should be no different. Attorneys should continue to take notes and connect with each other to discuss stipulations, deposition etiquette, how to handle objections, and the use of the deposition at trial. These conversations should also involve settlement discussions. In this way, remote depositions can help us build relationships with opposing counsel.

It is always important that a practicing attorney in a remote deposition appear in professional business attire, appropriate for depositions or hearing as if he or she were before a jury. While on the record, the attorney should make a very clear record, speak directly into the microphone and look directly into the camera.

An attorney who can navigate the complexities of remote deposition technology will always look more impressive to both the court and the jury. An attorney who can present exhibits with ease will always look better to the court and to the jury as well. Remote depositions should be carried out with a high level of preparation and decorum so when the judge or the jury sees them played back, they will provide an admirable and professional picture.

We hope these practical suggestions for the mastery of the technology and strategies for effectively taking and defending remote depositions are helpful. We welcome your comments and look forward to the evolution of remote litigation, which we and the Online Courtroom Project believe will be here to stay.

Chapter 11

Online Arguments to the Court

Theresa Wardon Benz

Before the COVID-19 pandemic, lawyers rarely made virtual arguments to a court. There were, of course, the occasional telephonic hearings, generally geared towards routine matters that could be resolved easily and without extensive argument. But most arguments to the court, whether a trial or appellate court, required actually going to court. Overnight, that changed because of the pandemic. In the early weeks and even months, some courts shut down completely, postponing hearings and oral arguments. But as the longevity of the situation set in, courts looked to the same video technology that allowed us to connect with family and friends, meet with colleagues, and even attend school. Appellate courts started setting oral argument by Zoom, Webex, or other platforms. Trial courts followed suit. Within months, it was not unusual to appear in several different courts—sometimes hundreds or thousands of miles apart—in a single day, and all without a plane ticket, or even shoes. Lawyers, judges, and court staff learned to adapt to this new reality.

There are upsides to virtual hearings and arguments, including ease in scheduling, reduction in travel, and open public access. But there are downsides too. It feels less momentous to appear in a state supreme court or federal circuit court by Zoom (and for U.S. Supreme Court practitioners, by speakerphone). It can be harder to "read the room." Even finding a place where you can stand, as you normally would, but be properly framed on the video screen, is challenging. Ultimately though, online arguments may continue post-pandemic. Perhaps they will not be the norm, but instead an option when a litigant cannot travel or wishes to reduce expenses. This chapter explores best practices for appearing virtually and making effective arguments to the courts. Fortunately, many of the steps lawyers previously took to prepare for arguments to the court, particularly appellate oral argument, still apply in the virtual world. Certain tweaks, however, can make the difference when it comes to making, and winning, your case.

First Step: Set Up Your Courtroom or "Studio"

The first challenge in effectively arguing in a virtual court is finding a physical place to do it. Your preferred courtroom or "studio" space may vary by type or argument. For a relatively routine discovery motion, you may elect to work and appear from home. But for a major dispositive motion or appellate argument, the office may be more suitable. Below are a few guidelines in choosing and setting up your space. Many of these points are covered in greater detail in Chapter 12 on communications, but I emphasize those most pertinent to arguments to the court.

A Strong Internet Connection

The last thing you want during oral argument is to have your video freeze or otherwise lose connection to the court. A strong internet connection can help avoid that situation. In assessing the strength of your connection, particularly at home, consider whether others will be using the internet at that time. If you live in a condominium building, for instance, your connection may slow when others are working. Or if you have school-aged children who might go virtual (or are streaming videos that use bandwidth), the same could occur. As with all things virtual, it pays to plan ahead and practice.

Free of Distractions, Including Your Background

A second requirement for your courtroom space or studio should be the absence of distractions. That begins with your background. Is the bookshelf behind you messy? Does it contain books or objects that will distract the court from your argument? The ideal background has a few objects for framing—perhaps a plant or piece of art, but is otherwise tidy and unremarkable. Your goal is to keep the focus on you and your argument. You can explore the use of virtual backgrounds, but they can appear contrived and there is a risk you will turn into a floating head, both of which would detract from your argument.

You will also want a space that is free, to the extent possible, of pet or human cameos, whether auditory or visual. If you will be conducting oral argument from your home, let others know not to disturb you and to refrain from loud noises or music during your argument. If you cannot keep your pet from barking or scratching a door, consider moving to an office space.

You also want to be free from distractions while conducting oral argument. Ensure your email is closed, so emails do not pop up during argument. Put your phone away, unless that is the method you choose for note-passing.

Ultimately, your goal is to create an environment that puts you at ease and allows the court to focus on the persuasiveness of your argument.

A Place to Present Where You Can Be Properly Lit, Framed, and Heard

Next, you will want to consider how you want to present. Are you more comfortable sitting or standing? If the latter, will you stand for the entire argument session, or do you want a chair nearby for sitting during opposing counsel's argument? While comfort is an important concern in choosing your presentation position, you should also consider what position psychologically prepares you for oral argument. Since most lawyers stand for argument in court, standing is more likely to make you feel engaged, as if you were actually in court. Additionally, you are less likely to slouch, swivel in your chair, or otherwise do things that are going to distract the court from your argument.

For these reasons, I could not fathom sitting during oral argument, so I worked with my firm to create a space where I could stand. It was difficult to position a camera and screen in front of a lectern with proper framing, so we used an adjustable standing desk. I was able to adjust the height for proper framing and comfort. I put my outline on a mini lectern placed on the platform. For lengthy arguments, I place a high stool behind me so that I can seamlessly sit on it when the opposing lawyer is making argument.

Whatever you choose, you should practice giving your arguments with your desired setup. It will allow you to work out issues such as proper framing and ensure you have your notes and outline in the right place. Practice the physical mechanics for your argument until they are second nature, much like being physically in court. Then you can focus on the substance of your argument and persuading the court.

In addition to proper framing, you will want to be properly lit. Shadows across your face or changing light are another distraction, both for the court and you. Ensure your lighting is planned and consistent.

You cannot be persuasive if the court cannot hear you or if you cannot hear the court's questions. Make sure your microphone is working well for your space and is set at the appropriate volume. The same is true for your speakers. And get familiar with how to use your mute button, so that you can mute and unmute with ease. The goal is to minimize distraction and stress, so that you stay focused on your argument and the court does too.

Finally, you want to choose a space that psychologically puts you in the courtroom mindset. There is a shift that happens when we step inside a courtroom. Create a space that allows for the same shift in your virtual argument. If your office or home has an extra room, a separate "studio" room may help facilitate this transition into courtroom mode. If a separate room is not feasible, setting up a standing desk and wearing a suit to your moots might help you feel game ready.

Second Step: Learn and Practice with the Technology

Courts are employing a number of different video platforms for argument to the court. While they all have relatively similar features, they are not the same. Thus, it is important to figure out what platform your court is using, learn its features, and practice your argument with it. Your framing and lighting may look entirely different using one platform over another. Likewise, what you see on your screen may appear different. To avoid surprises, practice with the same platform.

Another benefit of practicing your argument on the platform is that you can figure out any technological glitches or incompatibilities before oral argument. For instance, I discovered that a freestanding web camera was incompatible with a certain video platform, causing the video to freeze about ten minutes into the argument. The same issue did not occur with my laptop camera. It was only through practice that I discovered this. Accordingly, I was ready with my laptop properly positioned come oral argument time.

You should also practice selecting the correct camera and microphone in case they are not automatically selected when you log in. Knowledge of how to mute or share your screen is critical. Just as in a courtroom argument, ease in using technology and navigating the platform translates into an appearance of competence and calm in your argument. If you become frazzled because you cannot get your microphone to work or figure out how to unmute, it will be readily apparent to the court. It may also throw you off your game. Practice so that any glitches are dealt with quickly and calmly. If your office has information technology (IT) support, it is a good idea to have someone on standby during your argument.

Consider also what appears when you turn your video off. It is customary in some courts to turn your camera off until your case is called or while another lawyer is arguing. If you are using an office account, it likely will default to a professional photograph. But if you are using Google Meet or similar technology more likely to link to a personal account, you may, as I did, find a wedding photo pop up.

Additionally, if you decide to use backgrounds or other features, make sure you know how to use them properly. You do not want to show up in a court of appeals on a beach or a golf course. And no one will ever forget the lawyer who showed up to court as a cat.[1]

Many courts offer test runs, so you can ensure everything is working properly. I highly recommend doing those test runs and remembering the settings you used so that if it worked then, it will work again at argument.

Finally, you should capitalize on the fact that video technology makes organizing moots and other practice sessions easier and less time-consuming on your moot judges. This may allow you to do more practice sessions to refine your argument and presentation and get feedback from a wide variety of people, none of whom have to travel to participate.

Third Step: Learn and Practice New Protocols and Judicial Preferences

Virtual arguments have also introduced new courtroom protocols and preferences. Like you would in a courtroom argument, you should find out what those protocols and preferences are and practice your argument with them in mind.

For instance, in an appellate argument, you will want to know whether you are keeping your own time or whether you can see a clock on the screen.

You will also want to explore the view of the court, whether a single judge or an appellate panel, so that you can position the image in a way that allows you to make virtual "eye contact." This may require playing around with your monitors, laptop, and camera to find the best view for this. It should also allow you to see the judge and opposing counsel's facial expressions. This will make it easier to read the room.

Additionally, if you are using documents in your presentation, you should figure out how large those documents will appear both on your screen and the court's screen. If you are making a point with a document, you do not want it so small or unreadable that it becomes useless to the court. By the same token, if your image shrinks and goes to the side when you show a document, you will want to take that document down after you have made your point, so the court remains focused on you.

---

1. Daniel Victor, *"I'm Not a Cat," Says Lawyer Having Zoom Difficulties*, N.Y. Times, May 6, 2021, https://www.nytimes.com/2021/02/09/style/cat-lawyer-zoom.html.

Another protocol to determine in advance is how the judges will question. Will they take turns in a certain order (as in the U.S. Supreme Court) or ask questions as they arise? If the latter, some courts have a judge raise his or her hand. You may want to practice spotting raised hands and calling on judges—it feels quite unnatural at first. It is also helpful to practice your pacing and ensure you are giving adequate pause between points to allow judges to jump in with questions. The video format makes it harder for judges to jump in without those pauses. You may miss the cues on the video screen.

If the court does not issue a protocol setting out the method in which questioning will proceed, watch some previous arguments, if available. That will give you a preview of what to expect and some examples of what works and what does not.

Courts often insist that counsel mute themselves when someone else is speaking. This is something else that can and should be practiced. The last thing you want to hear from a judge is "you're on mute," especially if you just made a great point and your clock is ticking.

Fourth Step: Figure Out Where to Put Your Materials and How to Communicate with Co-Counsel or Client

The next obstacle is to figure out where to put your stuff and how to communicate with co-counsel. Your usual methods are unlikely to work in this virtual environment. For your notes and outlines, determine whether you can place them out of the view of the camera, but still close enough that you do not have to look down to see them. Remember that any move is exaggerated on camera, so a downward glance to your desk will be more pronounced than if you were arguing from counsel table and did the same thing. I tape my opener to the camera tripod, so that I can see it while making "eye contact." Some lawyers put their outline on the monitor and use it almost as a teleprompter. That can work well, so long as you can keep eye contact with the camera. Beware, however, of using your monitor as an actual teleprompter and treating your outline like a script. It will be obvious that is what you are doing and all the pitfalls of reading, rather than arguing—such as misplaced eye contact, increased pace, and lack of variability in tone—will manifest. Whatever you decide to do, practice is key.

You will also want to consider where to put your backup materials—the cases or record you might want to consult to answer a tough question or before your argument starts. If you prefer electronic materials, ensure you have access while your video argument is ongoing. If they are in paper, have them organized so you can easily find what you need without creating a distraction or moving out of the frame. Once again, planning and practice are key to making everything effortless.

Turning to your co-counsel or client, consider how they may want or need to participate. Most platforms accommodate observers (although some courts require observers to watch on YouTube or another livestreaming service, so that they cannot participate in argument). If the livestream is on a delay, consider that delay if your co-counsel or client plans to provide feedback or comments during the questioning. The better approach, if everyone is comfortable, may be to have your co-counsel in the room with you. Discuss how he or she will pass notes and how he or she will filter information from others. If your co-counsel is remote, have a plan for conveying important information, whether by text or email. As always, you are trying to minimize distractions to the court and distractions for you.

Fifth Step: Plan and Prepare as You Normally Would with All of the Above in Mind

This step is still the most critical. Most of what makes a good persuasive argument to the court pre-pandemic still makes a good persuasive argument to the court by video. You should know your record and the legal cases, have an opener that tells the court why you should win, and be prepared on a wide array of questions. None of that changes by video presentation. What does change, however, are the logistics and potential pitfalls. You will want to practice those, so they are second nature and you can focus on your argument.

Additionally, beware of the appearance of informality in video argument. It is easy to be lulled into it when you, opposing counsel, and the judge may be appearing from home.

Final Step: If Something Happens, Relax or Employ Your Backup Plan

Your internet may be slow. Your camera might freeze. Your laptop might lose battery. You might drop the binder you have perched precipitously on several treatises so the angle is just right. The dog may bark. Your spouse or child might walk in. There may be a siren blaring. Anything could happen. If something happens, take a deep breath and keep going with your argument. If it is a technological issue, employ your backup plan.

For example, you may have planned to use a webcam and monitor system, only to have it freeze mid-argument. Have a laptop ready to pop open and put into use if that happens. Or a tablet or iPad. Or your phone. Or all three. Make sure the video platform is already downloaded and ready to go. Courts are generally understanding and will stop your clock to allow you to fix an issue. Being prepared allows you to fix it quickly, which will allow you to stay calm.

Conclusion

The jury is still out on whether some virtual arguments are here to stay. Some appellate courts have already resumed arguments in person, with allowances for counsel or a judge to appear remotely when warranted. Trial courts may use video more frequently, particularly for less substantive matters. It is easier to schedule an argument when counsel does not have to factor in travel. Fortunately, some of the lessons lawyers have learned in this virtual world will carry over to in-person arguments.

Chapter 12

Credibility and Communication in an Online Environment

Lisa L. DeCaro

At the turn of the last century, when moving pictures created a whole new storytelling medium, the act of portraying the emotional truth of those stories began to change. Suddenly, the full-body expression of sorrow was both too much and not enough: the movements were too big and melodramatic, and the new medium required facial expressions not visible on large stages.

The goal of the actor is the same—to tell the story as clearly and honestly as possible—but the techniques required to do that were suddenly different. Both of those storytelling media still exist, and most actors endeavor to master the techniques necessary to portray a character with honesty and credibility in both.

Trial lawyers are now faced with the same expanded world—they must be skilled in presenting in a credible and persuasive way both in the courtroom and via video screen. Fortunately, actors have paved the way—they know what skills are required in each medium, and how to employ them. There are several fundamental differences between advocacy in a live courtroom and advocacy on a screen:

1. **Eye contact**. In a live courtroom, not only does eye contact work directly from the speaker to the person being spoken to, but it is also possible for everyone in the room to see who you are talking to. On camera, the audience receives eye contact from the speaker without the speaker actually looking at the audience. It is a difficult skill to master, but it absolutely is possible to build a connection with your audience through a video screen (just look at Walter Cronkite!).
2. **Body language**. Gestures that are effective when delivered in a large room, 20 feet away from your audience, become much too

large and distracting when delivered on a video screen, just a foot or two away from your camera.

3. **Voice**. When you are speaking to a jury across the room, your voice must be projected powerfully enough to be heard clearly throughout the space. On camera, you can modulate your volume all the way down to a near-whisper, provided your microphone is up to the task. Your options are limitless.
4. **The most important difference is this**. In an in-person setting, the *speaker* has an easier time connecting with their audience. On camera, the speaker has a harder job, but the *audience* often finds it easier to understand and evaluate the speaker. *The camera allows for a more intimate conversation with judge or juror, once the lawyer learns how to use the medium.*

Whether arguing in person or via video, the presentation "rules" for an advocate are, at the core, pretty simple: anything that distracts from your message is bad.

Eye Contact

This skill can be one of the hardest to master, and it can feel counterintuitive. But if we remember that your presentation is not about your comfort level, but rather about your audience's experience of your message, then the reason for learning this skill is clear. Your audience has a lifetime of experience receiving information via a presenter looking into a camera lens. They accept the convention immediately, and *they do notice* when a speaker is unable to accomplish it.

One benefit of online eye contact that tends to go uncredited: when you are making eye contact through the camera lens, you are actually able to make effective eye contact with every single person watching your argument. Every juror is able to feel equally connected to you and your argument, if you give them the opportunity by looking into the lens. There is some research (albeit nothing very recent, as far as I know) that suggests that triers of fact may feel less empathy toward witnesses who appear via video, as opposed to those who appear live.[1] What remains to be studied, however, is whether this holds true when the witness makes

1. Susan A. Bandes & Neal Feigenson, *Virtual Trials: Necessity, Invention, and the Evolution of the Courtroom*, 68 Buff. L. Rev. 1275 (2020).

effective eye contact through the camera lens, or whether the phenomenon occurs simply because the lack of eye contact in general can generate less empathy from a viewer.

Many people try to find ways to make it *look like* they are looking into the camera when in fact they are looking at their notes, or the video of an audience member. The reality is that only looking directly through the camera actually feels *to the audience* like eye contact. There is no effective substitute for looking straight into the camera. Mastering this skill requires practice (a *lot* of practice), but it reaps exponential rewards. It is important to figure out exactly where your lens is: On some laptops, for example, the lens is at the top of the screen, but that is not true of all laptops. And some laptops have a little green light next to the lens, but others have no light at all to show where the lens is. Find your camera, and look directly into the lens when you are speaking.

The technique works like this: When someone else is speaking, look at their image on the screen (not at the camera). When you are speaking, talk to your audience *through* the camera lens, while "listening" to them with your peripheral vision. In summary:

- Look into the camera—nothing else actually looks like "eye contact."
- Who (specifically) are you talking to? Have a clear picture in your head and talk to them through the camera lens.
- While someone else it is talking, look at their video.
- While you are talking, "listen" with your peripheral vision.

To make this easier, we recommend using an external camera, on either a tripod or a "boom" that can be easily adjusted to make sure that the camera lens is at eye level, in front of or directly above the monitor displaying the trier(s) of fact. This allows you to see the reactions of your listeners in your peripheral vision, to look at them while they are speaking without turning your head dramatically, and makes it easier to remember to look directly into the camera. Internal cameras on some computers are surprisingly high-quality and can provide a good picture. But they do not generally allow you to control the frame, contrast, white balance, and so on, and they can sometimes be harder to get to an eye-level height. If you are using your laptop camera, do not forget to clean the lens now and then. Grubby fingerprints do not enhance your appearance.

In this photo, the witness is looking directly into the camera lens.[2]

Figure 12.1

In this photo, the witness is looking slightly below the camera lens, at the attorney who is asking her questions on cross-examination.

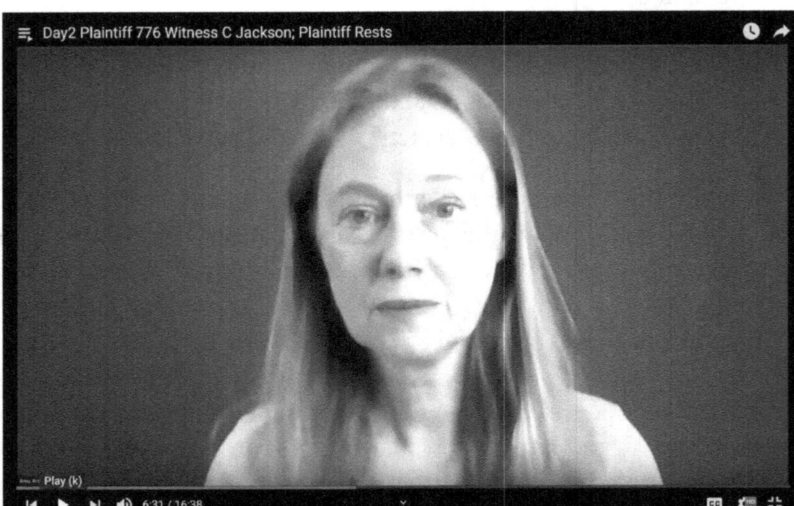

Figure 12.2

2. Screenshots from video of witness testimony (portrayed by an actor) in the Online Courtroom Project demonstration online trial, July 2020. Photo is the property of the Online Courtroom Project. Online Courtroom Project, *Demonstration Trial*, https://www.onlinecourtroom.org/demonstrationtrial (last visited Nov. 6, 2021).

Body Language and Nonverbal Communication: "The Camera Will Read Your Mind"

In *The Last Picture Show*, Ellen Burstyn had one scene in which she had to go through a variety of emotions—and express each one for the audience—without any dialogue. She asked the director, Peter Bogdanovich, how she was going to do that without any lines. Bogdanovich told her, "[J]ust think the thoughts of the character, and the camera will read your mind."[3]

To take this into the courtroom, ask yourself what your thoughts are revealing about you (and, therefore, your case). We have worked with many trial lawyers who have developed the habit of *acting* their suspicion of an answer, or distrust of a witness, in the belief that the trier of fact would be influenced by their opinions on the matter. While this is almost never accurate—this sort of "acting" does nothing but diminish the lawyer's credibility—it is especially dangerous to behave this way in an online environment. In a recent fully online jury trial, the judge and attorneys spoke with the jury after they rendered their verdict. One of the attorneys had been particularly aggressive with one witness, and during the post-verdict interview he admitted that he had been doing it somewhat "performatively" to make a point. One of the jurors commented:

> I see "performative" elements of this being really old school, and it makes me lose trust because I feel like I'm being performed to or entertained in an unnecessary way . . . we are all humans, and. . . . It just seems silly to me. Just making that connection, because we are over Zoom, is a little bit helpful.[4]

One of the less diplomatic jurors commented: "You really shouldn't be that d**k-ish on camera."

Gestures

Any movement appears larger in a small frame. Shifting from foot to foot, swiveling in your chair, waving your arms without purpose, and so on. All these common presentation problems—often just a sign of

3. Chris Nashawaty, The Last Picture Show*: An Oral History*, ENT. WKLY., Apr. 17, 2017 (interview with the cast and director), https://ew.com/movies/2017/04/17/the-last-picture-show-an-oral-history.

4. Interview with Online Jury Post-Trial, King County Superior Court (Mar. 18, 2021) (Juror 9).

nerves—can become a significant distraction on camera. Purposeless movement will at the very least distract your audience from your message. Far worse, it can reduce your credibility, and your ability to build a positive relationship with your audience.

At the same time, gestures play an important role in human communication (especially human communication that is supposed to be persuasive). To use your hands on camera, in a way that complements your argument, pay close attention to two important rules:

1 **Ensure that your hands are visible on camera**. If you are gesturing wildly below the camera frame, it will be distracting (at best!).
2. **Keep your gestures specific and intentional**. The general arm-waving that many of us are guilty of when speaking passionately can look completely insane on camera. Use your gestures to make points clearly (i.e., holding up four fingers while saying, "Ms. Jones called the plaintiff four times") or to emphasize key points. To make sure your gestures are not distracting, watch a recording of yourself practicing your presentation, but watch it with the sound off. This will allow you to focus on the ways in which your body language might be getting in your way.

Sitting or Standing

The physical elements of persuasiveness—energy, gestures, even our vocal inflection—are affected by whether we are presenting while sitting or standing. Both can be appropriate online, with special considerations for each.

Advocating to a judge or jury without standing feels fundamentally wrong to most trial lawyers (and to many judges). Presenting your argument while standing is an important way to bring back some of the formality and confidence you experience during an in-person argument. But it also creates some unique complications that must be tackled and practiced.

- **Framing**. When you are standing, you still need the camera lens to be at eye level, which means raising the monitors and the camera. Ideally, you want the frame to show you from approximately the waist up—it should show the top of your desk or lectern, to just over the top of your head. Ensure that the top of your head shows, but that there is very little additional "headroom" above your head. If your ceiling is visible in the shot, the camera is pointing up at you and should be raised to eye level.

Figure 12.3
Source: J. Bullinger, Persuasion Strategies

- **Sound**. When you stand, you will generally end up farther away from your internal or desktop microphone. A wired lavalier microphone can be attached to your lapel, and will result in better sound quality, and less extraneous noise picked up from your notes or other motion in the room. In addition, a mic with a physical mute button will make things easier, since you will not need to step forward to mute and unmute your microphone within the online platform.
- When you are not speaking for a long time (during the arguments of other parties, etc.), you may decide that you want to sit, but do not want to worry about reframing the shot when you stand to deliver your argument. Using a tall stool can be helpful in this instance—you can sit in the same frame as you occupy while standing, with minimal adjustments to the camera shot.

More casual presentations (or elements of trial like direct examination, where the focus should be on the witness rather than on you) might be better done while sitting. But do not make the mistake of assuming that this position does not require the same attention to detail and practice.

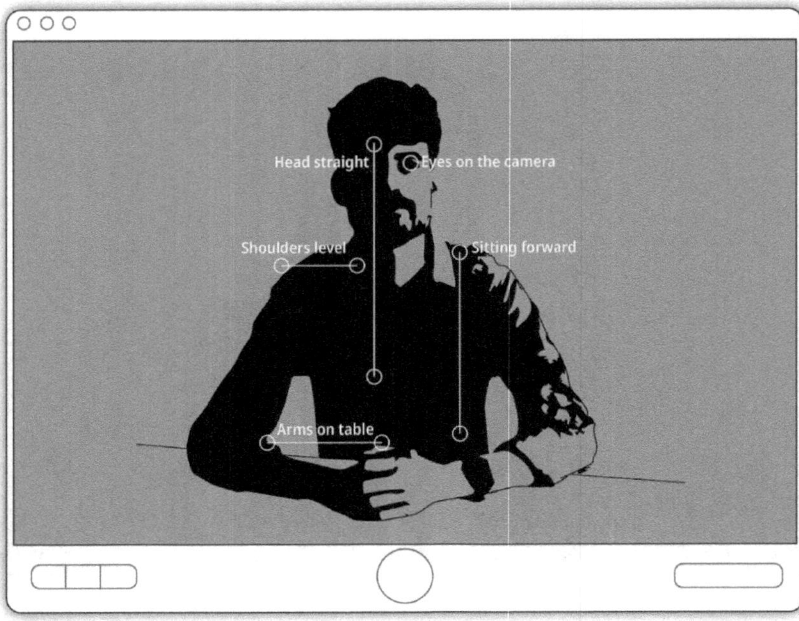

Figure 12.4
Source: J. Bullinger, Persuasion Strategies

- **Framing**. Contrary to what we have all become used to seeing on video conferences, the most effective frame is not one that encompasses your ceiling, your head, and a bit of your neck. Look at the frame used by television news anchors—they fill the frame, from the top of their head to the top of their desks (where their hands are visible, usually resting on their notes).
- **Sitting position**. The position held by the news anchor—a forward lean, hands on the desk in front of them, straight back—has been consistently rated as more credible than any other sitting position. It projects a sense of confidence, openness, and helpfulness. For this reason, strive for this position on camera, and teach your witness to do the same. To make it easier to maintain this position for an extended period of time, and to prevent yourself from absentmindedly sitting back in your chair, push your chair back so that you are perched on the front of your seat. This tilts your pelvis forward, mimicking an "ergonomic" backless chair, and will make it possible to hold the position without fatigue in your back.

Voice

Those of us who are used to public speaking have spent a lot of time learning to *project* our voice, to be heard clearly and distinctly by ears scattered around the room, sometimes quite a distance from us. Online, the goal is the same: a dynamic, pleasant, confident, clear voice. But the steps to achieve that goal are different. You will, by necessity, be heard through a microphone, which has both advantages and hidden pitfalls:

- A good microphone is essential. In our discussions with jurors and mock jurors over the past year, it has become clear that sound problems are even harder for the trier of fact to overlook than are video problems. Built-in microphones have several inherent drawbacks: they are closer to your keyboard and laptop fan than they are to your vocal cords, they pick up everything else in the room, and they limit your ability to stand or sit farther away from your computer to achieve the frame you want. Use a quality external microphone, and choose the right type of microphone for your room type, presentation style, and equipment setup. (See Chapter 6 for a more detailed discussion of microphones.)
- A room with carpeting and soft furnishings is generally going to give you better sound than one with hard floors and surfaces, which can become like an echo chamber, bouncing sound around the room. Any echo or bouncing sound will make it harder for your audience to understand you.
- In any presentation, it is important to speak more slowly and clearly than we do in "real life." This is especially true when presenting on camera, since rapid speech patterns or mumbling can cause your audience to get lost, bored, or feel overwhelmed. Slow down, speak intentionally and clearly, and focus on the message being received by your audience.
- Work within the limits of the technology—if you interrupt someone, they will not hear the beginning of your statement, because it takes some time for the tech to shift to you as the speaker. Similarly, if opposing counsel objects during your presentation, or a judge wants to interrupt with a question, you are apt to miss the beginning of their statement. It is another reason to speak more slowly and clearly than you might ordinarily, and to keep "listening" to the screen with your peripheral vision while you are speaking.

Unless you are trained otherwise, most people will tie their volume to their level of energy when speaking. In other words, when you are confident, passionate, and committed to a statement, your volume is naturally higher, and your rate of speech is faster. For this reason, trial lawyers are consistently rated as more credible when they are projecting their voice at a higher volume. But using the volume that makes you look confident and committed in a large courtroom, when you are speaking to jurors through a camera a mere foot or two away from you, will seem overly broad, overly aggressive, and even cartoonish. As film acting had to evolve to become more "natural," so too must you modulate your volume and pace for the microphone. However, be careful that while you might lower your volume, you should not lose the energy and intensity you need your listeners to feel from you. Like with most elements of effective communication, if you focus on your listener the problem is likely to resolve itself. Remember who you are talking to, where they are in relation to your voice (very close, if you have a good microphone), and what they need to hear and see.

Setup

You have many options in terms of the equipment you decide to use, and the way you set it up. Much of it will depend upon your available space, how often you intend to use the space, your budget, current technology, and other factors. See Chapter 6 for details on the types of equipment you might want to investigate. Whatever you decide to try, be prepared for a bit of trial and error. It takes some time to get your space exactly right, and probably more than one trip to the store to exchange one thing for another.

And it matters. As one juror in a recent online trial commented:

> The plaintiffs' attorney had high quality video and audio and he had a ring light and he came off crisp and clear and the defense attorney looked like he was using the camera on his laptop so he was just sitting at his desk. The lighting was not good, the picture was not sharp, and in focus. He even lost connection from his home and had to drive into his office to be able to connect there. But he was still using the same crappy laptop and camera at his office. He's a lawyer for crying out loud. I don't have a sense this affected the verdict. [However, the jury did reach a verdict for the plaintiff.][5]

5. Interview with Kevin Burkhart, Presiding Juror in Online Civil Jury Trial in Multnomah County, Oregon (March 2021).

There are some general tips to keep in mind, related to communicating with a judge or jury:

- **Camera**. Using an external camera allows you to control the angle, frame, and appearance of your picture more precisely than you can with an internal camera on your laptop. One important note, though: most webcams are designed with a wide-angle lens, to show an entire conference room. Using one of those webcams will make it impossible to frame yourself properly and will distort your image (like looking through a fish-eye lens). Choose an external camera that has a zoom feature, which can be adjusted to frame you properly. If you do not have an external camera, find a way to put your laptop camera at eye level, and try to achieve the best frame you can. You might use a standing desk, or even a stack of books to raise the laptop into the best position. Be creative.
- **Microphone**. The microphone you choose will depend on the rest of your setup, your room, and whether you intend to sit or stand while presenting (see Chapter 6 for more details). Make sure that your microphone has a mute button, that it picks up your voice clearly without a lot of background noise, and that it is easy to use. Beware of a wireless microphone, as tempting as it may be: if something causes it to feed back or cut in and out it can be worse than nothing at all. I generally suggest a wired lavalier microphone with a mute feature.
- **Background**. A cluttered background will become a distraction from your argument. On the other hand, a plain white wall makes it look like you are posing for a passport photo—or a mug shot.
    - If you are designing a space that will not need to be broken down after every online session, consider a carefully curated bookcase, or a desk with a simple piece of abstract art above it. A plant or two is a great way to warm up a space without it becoming too distracting. A synthetic plant can look great on camera, and its appearance will remain consistent.
    - If you do not have the luxury of setting up a dedicated space, you can eliminate a distracting background with something as simple as a sheet (a solid pastel color will look best on camera) hung as a backdrop. This may be especially important for your witnesses, who may not need a permanent space, and may be appearing from their bedroom or cluttered office.
    - Avoid a "virtual background" at all costs. At best, they will make your audience wonder what you are hiding. At worst, they will fade in and out and cause you to look like

a floating head. At one meeting recently, one lawyer had a great-looking virtual background, which looked real. Until her husband handed her a cup of coffee, which appeared like a floating hand and cup into the frame.

  - Check your image to make sure nothing behind you is distracting (like a lamp that appears to be growing out of your head) or private (like confidential client files relating to another case).

- **Lighting**. From the perspective of persuasive communication, the goal with your lighting is to allow your audience to see you clearly, including your facial expressions. Lighting can make you look younger by washing you out, but it also eliminates one of your most important tools of persuasion—your facial expressions. Do not sacrifice persuasion for vanity.
  - Everyone looks better with front, *diffuse* lighting. A bright window in front of you can be good sometimes, but light moves during the day so I recommend lighting in front of you to supplement the natural light, even if you have a big window in the perfect position.
  - The ubiquitous ring light can overexpose your features, making you look washed out. It can also reflect off your glasses, or even your eyes, giving your audience an image reminiscent of the horror movie of the same name. Generally, a photography "box light" will provide more diffuse light without as many concerns about reflection. Keep light sources above the camera, rather than at eye level, to avoid reflection off your glasses or pupils. (I have included some other tips for reducing reflection in the discussion on eyeglasses, below.) You want the light source to be slightly higher than your head, but not so high that it creates shadows under your eyes.
  - In addition to lighting from the front, consider using lights to create dimension in your background. A back ("rim" or "hair") light will brighten your back, shoulders, and hair, helping to keep you distinct from your background. A "background" light can be focused specifically to illuminate the wall behind you, which can make your setting look more professional, warm, and appealing. With any additional lighting you choose to use, ensure that no light is shining directly into the camera. This can be a problem with a window or natural light source that is not directly in front of you. A light source shining into the camera will cause the camera to adjust to that light, and will leave you as a darkened silhouette.

Figure 12.5 The online studio at Wheeler Trigg O'Donnell, LLP
Source: Laurie Dittbrenner

Other Considerations

Notes

Where will you put your notes? Practice various things until you find what works for you. The important thing to remember is this: just like in person, you do not want to try to sneak glances at your notes. You will not fool anyone anyway—they know when you are reading a script or checking an outline, and they do not think less of you for it as long as you do it unapologetically. Stop speaking, look at your notes, then reconnect with the camera lens and resume. Sneaking glances can make you look shifty—and remember that above all, you need to be *credible.*

If you need to read a scripted presentation, rather than just checking an outline, you do want to put the text as close to your camera as possible, remembering that you may also need to be able to see your audience. For scripted presentations, some options that have worked well for my clients are:

- Text on a large screen, with the camera on a tripod in front of that screen. This works if you are giving a presentation to one person, or a webinar in which you do not need to see your audience. If you need to be able to see your audience on that screen with your peripheral vision, it can be hard to do so with your notes sharing that screen.
- If you are presenting with a PowerPoint slide deck, put the portion of the script pertaining to a given slide in the "Notes" section of PowerPoint. Start the slide show and use "Presenter Mode" to allow you to see your notes on your screen, while only sharing the show itself with your audience.
- Teleprompter technology is available now, using an iPad with a special mirror to project the text in front of the camera lens. Clearly, this only works for scripted presentations, and it requires a lot of extra practice. Reading from a teleprompter and making it sound natural is a very specific skill. Using a teleprompter can also complicate things: you (or an assistant) must scroll the text at the appropriate speed, and you need to figure out what you will do if the teleprompter freezes or stops working.

A note about notes: this applies to the lawyer/presenter. Your witnesses should obviously not be using any notes other than the exhibits they are known to be using. (See Chapter 9.)

Assistance

It is very helpful to have someone else from the trial team in the room with you, as you might have someone with you at counsel table if you were there "live." Having someone else there to take notes, help you fix any tech problems, and allow you to focus only on communicating your message clearly to your audience, can be priceless. At the very least, co-counsel or a paralegal should watch the jurors while you are speaking to them through the camera lens. One advantage to the online system is that the jurors are more relaxed, and less guarded, than they generally are in a courtroom, allowing an observer to see honest reactions to what you are saying, as well as glimpses into their home and life.

Clothes

The most common question I get (right after "where should I be looking?") is "what should I wear?" It is an important question, and people are right to ask it. Those who do not give this question enough thought can seriously damage their credibility, and their case. As with the other advice we have discussed in this chapter, your goal should be to present yourself professionally, and to wear nothing that distracts the trier of fact from your argument. In general, you should be dressed for the courtroom—even from the waist down, including your shoes. We have all heard about lawyers appearing before a judge wearing "business casual" (or far worse!), and many judges are complaining about it. The news—legal and otherwise—is full of reports of witnesses or lawyers appearing from bed, poolside (in a bikini), or hair salons, and with a glass of wine or a cigar in hand.[6]

The online environment can feel less formal. Although this can help the jurors to feel more comfortable sharing information during voir dire, the officers of the court still need to show respect to the judge, and to the process. In addition, the clothes and shoes you wear affect the way you carry yourself—do not carry yourself like someone who has not left the couch in ten months.

Do not forget to have this conversation with your witnesses, as well! Even those who would not show up in court wearing sweats and a

---

6. Debra Cassens Weiss, *Lawyers Smoke Cigars, Drink Wine during Zoom Hearings; Litigants Appear from Hair Salon or While Driving*, A.B.A. J., Feb. 16, 2021, https://www.abajournal.com/news/article/lawyers-smoke-a-cigar-drink-wine-during-zoom-hearings-litigants-appear-from-hair-salon-while-driving.

T-shirt do not necessarily know that the dress code still applies to online proceedings.

A few special considerations for clothing on camera:

- White can be overly bright on camera and can distort the camera's white balance feature, causing the other colors on screen to look overly beige or blue.
- Red can be harsh and jarring on camera and can actually appear to be fuzzy around its edges. Avoid very light and very dark colors—most people look best in pastels, like an oxford blue.
- Avoid tight patterns. Small stripes or patterns with high contrast can appear to move on camera, creating a very distracting effect (moiré effect).
- Jewelry, shiny buttons, and other metal can reflect light back into the camera. Try to keep the bling to a minimum.

Eyeglasses

If there is glare from your glasses that makes it impossible for the audience to see your eyes, you are negating the benefits of eye contact we discussed earlier in this chapter. But usually, removing your glasses is not the answer—if you cannot see the screen well, you are unable to "listen" to your audience while they are on screen. Here are some solutions to this problem:

- Order prescription glasses with glare-free lenses. Even readers can be ordered with this coating, from an optometrist.
- Anti-fog spray can help, although it does not eliminate glare entirely.
- Tilt the earpiece of your glasses up into your hair a fraction of an inch, causing your lenses to tilt downward. This will not be visible to the audience, but it can eliminate glare from your lenses.
- The best solution is to keep all of your light sources above the camera, rather than at eye level. That will keep your face lit, without reflecting off your eyes or glasses.

Final—but Vital—Tips

Practice, Practice, Practice. The importance of practice cannot be overstated. During your argument, your focus should be on connecting with your audience (through the camera lens), listening to your audience (with your peripheral vision), and clearly communicating your message.

You should not be worried about your equipment, technology, or the new skills required (like making eye contact with a camera). Practice early and often, until you are able to focus only on your argument.

- **Practice the entire process, not just your argument**. Practice with the system and platform you will be using for your argument, in the room where you will be doing it. The framing, lighting, control of what you are seeing on the screen, all can change from platform to platform.
- Check to make sure your **speakers** are not causing audio feedback. If they are, you may need to use headphones, or otherwise adjust your setup.
- **Practice** by yourself until you feel comfortable with the technology. Then practice with a trusted colleague or friend, to learn how to "listen" with your peripheral vision.
- **Turn off your self-view**, if the platform allows you to. We all have an annoying tendency to look at our own image on the screen, and fix our hair or suit accordingly. But if you are focusing on yourself, you are not focusing on your audience! And that kind of self-consciousness is a killer, whether you are presenting in person or online. You would not want to look at a mirror when you are making an in-person argument to a judge, so do not do the equivalent when making that argument through a camera lens.
- Once you are comfortable with the tech, practice your argument in this environment. Record a run-through, and then watch it with the sound off. This allows you to see what the shot looks like, check for distracting gestures, backgrounds, and so on, without focusing on how much you hate seeing yourself on camera. Then play a bit back with sound—only enough to make sure the sound quality is okay, and that your diction, rate of speech, and volume are appropriate. It is easy to record and review your practice sessions using most platforms, even if you are the only one on the meeting.
- Practice with more than one person on the conference, experiment with moving their images around (if you can within the platform you will be using), and determine what works best for you.

On the day of your argument, testimony, or presentation:

- Turn off notifications, close all apps, and restart your computer. Make sure the only applications you have running are essential for the presentation. You may not be able to turn every notification off, but do your best. Turn your cell phone off entirely, unless you need it to communicate with your team. If your phone rings through your laptop, turn that feature off if you need to keep your phone on.

- Do one final run-through. Tech that worked yesterday has a tendency to decide not to work today. Make sure the framing, lighting, sound, internet, and so on, are all working the way you want them to.
- Restart your computer again. While helping two separate lawyers prepare for online hearings, I discovered that the platform the court was using in that state had a strange quirk: if they used the platform after using a different platform earlier that day, *and* if they were using an external camera, the platform would crash 15–20 minutes into their argument. If they restarted their computer after using other platforms, this problem did not occur.
- Log in to the actual meeting as early as you can, to check to make sure everything is *still* working—and with enough time to troubleshoot if something is not working as planned. Check your appearance and framing, and make sure there are no wayward filters on! We all saw the unfortunate lawyer who had to inform the judge at his hearing "I'm here live. I'm not a cat."
- Turn off your self-view, and focus on your audience.

Always have a backup plan.

- What will you do if you lose your internet connection? Hot spot using your cell phone, or a hot spot device?
- What will you do if your computer crashes? Have a back-up computer or other device ready to go.
- What if you lose sound? Learn how to change your sound and video devices using the specific platform you will be using, so you can change to your internal camera or microphone if necessary.

Relax, and Accept That You Cannot Control for Everything

At some point, no matter how hard you have tried to anticipate everything, something will go wrong. Honestly, this is to be expected, and everybody understands. What matters is whether it is a problem you should have anticipated (like a lack of familiarity with the platform or your equipment), and how you respond to the problem. Your reaction will speak volumes about your confidence, your credibility, and your professionalism.

A longtime friend, who was a Queen's Counsel (QC) in Alberta before retiring, found early in his career that the best things that happened in the courtroom—the ones that helped the most to build a relationship with the triers of fact—were the mistakes. At his first jury trial, he spilled a glass of water all over his notes during his opening statement. Once the initial embarrassment wore off, he noted that the jurors were looking at him with a lot more kindness and much less skepticism. You cannot create these "happy accidents" on purpose, and trying to do so will decrease your credibility. But you can recognize that when they do happen, they are not the end of the world.

Do Not Forget Your Witnesses

Witness prep is essential—your witness will be under the same scrutiny you are under on camera, and is probably even less comfortable with the online environment than you are. Prepare them using the platform they will be using to testify, and help them to feel comfortable with the process. (See Chapter 9 for some valuable advice for client and witness meetings.)

Chapter 13

Implementing Technology for Remote Trials

Josh Splansky, Ted Brooks, Alicia Aquino, and Noah Wick

Introduction

Chapter 6 of this book addresses the motivation to address the technical issues of online court. This chapter, in contrast, focuses on an approach for executing it. Specific technology can become obsolete quickly in any emerging field. Therefore, the focus in this chapter is more on how to generally incorporate technology into online trial practice and only refers to specific technical baselines where prudent.

At the end of the past century, law firms and others were panicked about the impending doom of the world's computer systems expected to occur at the stroke of midnight—known as Y2K. Even the largest and most technologically advanced firms had an all hands-on-deck state of alert in case the world's computers all crashed to a standstill at once.

Twenty years later, we were brought to a similar point—this time, with COVID-19. When law firms were evacuated and the courts were all closed, we were faced with the unique challenges only a pandemic could present. At first, we were all willing to wait it out a couple months, then for just a few months longer. Eventually, this delay lasted into 2022 and beyond, before we once again see jury trials proceeding on a "normal" schedule.

The good news is that while this all seems new and different, we have successfully adapted to the point that remote proceedings are now a daily occurrence, with courts across the country making the transformation in less than a year. In this chapter, we will explore several technological adaptations to in-person jury trials, which, in fact, are not all that different from the "old fashioned" courtroom trial presentation technology we have been using successfully for the past 20 years.

Our goal is to create a guide for people to reference when addressing the technical needs of an online trial. The specific solutions are dependent on both the type of trial you have and the role you play in that trial.

Courtroom Setup Options: In Person, Remote, Hybrid

The Traditional In-Person Trial

We will not focus in great detail here, although the pre-COVID, live, in-person courtroom trial is the standard to apply when considering other options. Even so, in-person trials may be different for quite some time, given the addition of plexiglass dividers installed in many courtrooms for the judge, witnesses, and jurors. Current restrictions can also include mask-wearing mandates. Many courtrooms have a single projector screen or large monitor installed, which is not always practical to use in combination with greater social distancing requirements and more distant seating arrangements. An in-person trial can now require a significantly greater degree of technology, given that participants may be distributed all around the courtroom, including in the gallery. The jury box might hold only three or four jurors. Additional monitors will need to be placed such that everyone can easily view the evidence displayed. Sometimes, the gallery will be relocated to another room altogether, forcing additional technology to not only relay video of the attorney, witness, and judge, but also to retransmit the electronic documents to the gallery as well.

Remote Trial

While this is the most drastically different alternative to being in the courtroom, the remote trial offers a great deal of flexibility to litigants, and cost savings because travel expenses are greatly reduced or even eliminated. Two of the greatest concerns for attorneys have been losing the ability to look at jurors to judge their credibility, and the personal connection that comes from being there in the same room with everyone. Ongoing research is looking at the differences in remote and in-person communication, and we will not comment on that here. As to the lack of ability to judge the credibility of a remote witness (or juror), our practical experience is that, when compared to sitting 20 to 30 feet away from a potentially masked witness, being able to use a computer monitor to actually see the witness's eyes and facial expressions is certainly the better option.

Hybrid Trial

The hybrid trial has quickly become a popular option involving remote appearances and will likely remain so based on its convenience and cost savings when compared to traditional in-person options, which frequently involve travel expenses and scheduling concerns. However, the hybrid trial also requires a comparatively complex technology setup in the courtroom.

In a hybrid trial, people are physically present in the courtroom, including the judge and court staff, court reporter, some or all counsel, and possibly a jury. It should be noted that masks may be required inside the courtroom for some time to come. Some or all witnesses may appear remotely, in addition to counsel. Equipment setup must take into account the views that each person will have of all of the participants from both inside and outside the courtroom. While a remote trial would likely feature each attorney appearing individually using their own webcam, the view from inside the courtroom would be different. Rather than placing individual cameras on each attorney, it is recommended that a wide-angle camera be placed such that both counsel tables and their respective attorneys are included in one view. This is closer to what a witness would see if that witness were physically present on the witness stand. This camera ideally should be placed near a large display or screen, in order to allow counsel to appear to be looking at the remote witness, and vice versa. This arrangement helps to come close to replicating the experience of eye-to-eye contact, a critical aspect of human interaction and conversation. As noted in Chapter 12, to completely replicate the experience of eye-to-eye contact, attorneys and witnesses need to practice looking directly into the camera lens rather than generally at a monitor.

Another camera should focus on the judge, with yet another on the witness stand for in-person testimony. In this scenario, a total of three cameras is required.

Two additional important concerns are internet access and audio systems. In many cases, an existing conference call phone system may be utilized, along with microphones and speakers already in place. Other options include setting up a separate system, giving special consideration to the potential for echo and audio feedback issues.

Many courtrooms offer public Wi-Fi access, but in addition to minimal security, the signal is often spotty and unreliable. While a personal hot spot may work, an industrial-strength device is recommended. Although remote platforms have relatively low bandwidth requirements, data consumption can run two gigabytes (GB) or more per hour—quickly exceeding many "personal" data plan limits.

The remote platform may be hosted by the court, court information technology (IT) staff, or by a vendor as stipulated. Exhibit presentation may be provided by counsel or a trial technologist from inside the courtroom or remotely. To ensure maximum visibility with minimal distractions, we recommend that trial presentation software, such as TrialDirector or OnCue, be used, as opposed to a PDF viewer or embedding in a static slide show.

Personnel

For the Courts: Technical Bailiff—Implementation

A technical bailiff becomes the only novel addition to a courtroom. It is an extension of three traditional courtroom roles: court admin clerks, courthouse IT, and courtroom officers (bailiff).

More simply, a technical bailiff is responsible for hosting and administering the online session. He or she distributes the log in link for the session. The technical bailiff assigns each participant access to different breakout rooms—virtual chambers; gallery; jury room; waiting rooms for upcoming witnesses; counsel rooms—and moves the groups to their assigned room during each recess. This individual may also rename participants to help identify them to other participants. By communicating with people in the various breakout rooms, particularly to ensure the presence and readiness of the jurors and witnesses, the technical bailiff will also police the event for unwanted attendees and unmuted noise. The technical bailiff should also be able to provide simple troubleshooting, and should be able to be reached by attendees via a chat message or other means. Finally, the technical bailiff should be accessible to jurors during deliberations in the event that they have questions or experience any technical problems.

For Counsel: Hot Seat—Electronic Display of Evidence

Electronic display of evidence has been evolving in the courtroom for more than 30 years as the personal computer has matured. Sophisticated presentation software and document cameras have easily transitioned to online trials. Documents streamed in an online trial often have perfect resolution and appear on each individual's monitors as clearly as they would in the most modern courtrooms. In fact, the logistics of setting up (or supplementing) physical courtrooms for electronic display have been time-consuming and expensive. With online trials, the ability to display evidence is already a feature of the basic web conferencing platforms.

The question is not whether electronic display works in remote trials, but rather whether participants can master the subtle differences between its uses in the physical and virtual courtrooms. In the physical courtroom, the screens are one of several places that participants will look. It may resemble a tennis match, as people turn their heads back and forth between the judge, counsel, witness, and screen. If a document stays static on the screen too long, it simply becomes stale and fades into the background—as people follow the active testimony (or whatever else is worth watching—a defendant's reaction, someone snoozing in the gallery, or a paralegal scurrying to respond to a sudden request). In a virtual trial, electronic display of evidence through sharing of a screen often dominates most of the on-screen visual field, relegating the participants to a corner. Specific platforms might allow for further customization, but in most cases the shared screen will occupy a majority of the space. It is less a tennis match and more resembles looking through the windshield when driving a car: all of the action is directly in front of you. Visual real estate on the monitor is a zero-sum situation: if a document is full screen, the witness and everyone else are not. Knowing when to remove a document becomes almost as important as showing it. While some platforms allow for dual monitors, the presentation needs to cater to the most basic setup. Unless everyone has multiple screens, the expectation should be only one monitor.

With pixel real estate at a premium, the online courtroom requires a greater emphasis on clean display of documents. Trial presentation software that allows for just the presentation screen to be shared with the court is highly recommended. While it is usually possible to use a PDF viewer instead, it will lack some of the polish of a dedicated presentation software. Whatever the software preference, the shared view should only be of the individual program or a second (extended) desktop monitor. Sharing the primary desktop screen will reveal all the open tabs, desktop programs, or email notices. Such unintended display could have dire consequences. Present only from an approved device—preferably by someone other than the presenting attorney. Ideally, the person controlling the presentation should be someone who is very familiar with trial presentation software and with the specific web conferencing platform that is being used.

Cost of Doing Business

It is estimated there are some 20,000 courtrooms in the United States, with only about 10 percent of those currently equipped for trial presentation and remote technology.[1] Although a decent retrofit for a single courtroom can run upwards of $50,000, some courts have utilized funding available via the Coronavirus Aid, Relief, and Economic Security (CARES) Act or may be supplemented by budget contributions from

1. Personal Communication by Ted Brooks with Adam Lofredo, Director of Justice Division, Conference Technologies, Inc. (Apr. 6, 2021).

frequent users of the systems such as district attorneys and public defenders, or usage fees from civil litigants.

While the cost of a permanent installation falls upon the court's budget, temporary "one-trial" installations are available from many trial technology providers. These portable single-use installations and associated rental costs are typically covered directly by the litigants. This may present a significant hurdle in cases of lesser value, however.

Regardless of who is paying the bill—taxpayers or litigants—it is our recommendation that nonproprietary hardware and software are utilized whenever possible, thus avoiding early obsolescence, incompatibility, limited support options, and the inability to adopt and incorporate new technologies as they become available. Further, "off the shelf" systems allow for easier integration of additional equipment or upgrades that may be required by individual courts, such as Americans with Disabilities Act (ADA) listening devices, remote interpreters and court reporters, recording devices, closed sidebar audio systems, and additional cameras or monitors as needed.

This approach of using nonproprietary systems also helps promote a level of standardization from court to court. Rather than being forced to learn and use a unique platform for each individual courtroom, attorneys and litigants should be able to easily connect and use familiar platforms with currently available devices.

Most of the equipment needs are covered in detail in the section A Room with a View: A Studio Primer, below. But there are some basic needs that fall outside the dedicated studio setup. For one thing, a studio setup is only for those presenting during the trial. Many participants' role will largely be as a viewer. For these viewers, simple laptops or Chromebooks with integrated webcam may suffice. A second monitor may not be needed, but could still be useful for any participant.

A Room with a View: A Studio Primer

For online trials, the best setup is one in which each person accesses the trial individually. As soon as you begin accommodating more people and larger spaces, adequate audio and video becomes much more complicated and expensive to do correctly. Traditional courtrooms themselves are simply not built for video. They are typically very big with high ceilings making it very hard to mic and light.[2] They have participants in

2. Interview by Josh Splansky and Marc King with Jim McMillan, Principal Court Technology Consultant, National Center for State Courts (Apr. 22, 2021).

every space. Expanding a setup to correctly capture everyone in a traditional courtroom can quite literally require a television crew.

If everyone presenting in online proceedings needs an individual space from which to present, then every presenter needs some sort of "studio." The directive to find suitable space is straightforward: a room with good lighting, acoustics, and internet, with a suitable background and removed from personal distractions. This section provides a technical primer on how to turn that space into a trial-friendly studio, whether in a courthouse, an office, or a residence.

Location

The immediate expectation regarding where to locate the studio may simply come down to which room is available. Which room has a backdrop that could be on display to my colleagues? Which room has a door that I can shut for separation from others? Which room has the best light?

Before considering any of those factors, think internet. Which room has the best internet? Give strong preference to rooms with a wired connection. Failing that, ensure your Wi-Fi signal is as strong as possible. You may use your laptop or cell phone to view the Wi-Fi signal strength in each room you are considering. Unreliable internet will jeopardize the whole experience.

Pick a place where you control the internet and can control the environment.

Windows can be a nice source of natural light, if the light is in front of you. Natural light is good, but is inconsistent depending on the time of day, time of year, and the daily forecast. Most of all, make sure there are no windows or other lights behind you that are visible into the camera. The light will flood the image and webcams will auto-adjust to darken the image of you.

Camera

Imagine being content with your home television set. And then you walk into an electronics store and see something with higher resolution, or significantly better contrast or colors. You never even knew you cared about a better picture until you saw it firsthand.

Many laptops have built-in webcams. But almost all of those built-in cameras do not look as good as a dedicated webcam. Your webcam does not have to be top-of-the-line, but it should have at least a native resolution of 1920 × 1080 (often marketed as 1080p). Most current teleconferencing platforms do not stream in 4K (resolution of 3840 × 2160) or higher, so getting a 4K camera will not improve the resolution beyond

1080p. However, higher performance cameras (whether 1080p, 4K, or somewhere in between) may have other attributes that separate them from lesser options. Good cameras will have better color, as well as being better at maintaining high-quality autofocus. In imperfect light, a good camera will perform better. Good cameras avoid blur when capturing fast movements (important for those who talk with their hands). Adjustable field of view (FOV) or zoom features might also be features found in better quality cameras. FOV adjustment allows for you to put the camera in the exact right place, and then narrow the frame to adjust to you and your background. While most USB webcams are plug-and-play, you may still need to install their recommended software to access certain features. Avoid wide lens features as it offers no benefits and could introduce an unflattering fishbowl effect on you.

There are also other good options beyond webcams. You could opt for a conference room camera, as long as you can zoom to a properly framed shot with the camera at eye level. Higher end DSLR cameras, action cameras (GoPro), and digital camcorders can also look very clear, but there are a few additional concerns to address. These cameras have more features and can be more complicated to use. Make sure that you have a way to reliably connect these cameras to your computer, and that the camera's output is clean (or can be cropped to a clean image)—as some cameras display the image information digitally stamped on their output to your computer. You do not want the resolution settings, time code, or viewfinder lines visible. There are some fantastic cameras that were not designed for webcam usage, and even some very high-end cameras make it impossible for users to remove display information.

Camera placement can be directly on top of the laptop screen. When the laptop is properly positioned at eye level, presenting to the camera is straightforward while maintaining easy viewing for everything on the monitor. Some people have had success with creative placements so the camera sits in the line of sight between the viewer and the screen. This way, the presenter can look at the screen while simultaneously appearing to make eye contact with everyone. This is very tricky. When it is done well, the results can be very effective. But all too often this configuration only works when you are looking at the middle of the screen to mimic making direct eye contact (not, coincidentally, the portion that is obstructed by the camera). When you look to the edges of the screen, you can appear like someone who is *avoiding* eye contact, leaving the viewer with exactly the opposite of your desired effect. One might think to simply use a bigger screen and move it farther back (that way even a small camera can occupy a larger sweet spot. The possible problem here is that as you focus at the distant screen and not the nearer camera, you may lose the eye-contact effect. It may appear like you are looking past the person to whom you are talking (which makes sense, because you quite literally are looking past the camera).

Recognizing eye contact is engrained in humans. Teleconferencing may be an emerging skill, and even being in court is unfamiliar to many. But we are all experts with eye contact. "Almost eye contact" can be worse than no eye contact because we innately find it off-putting. Make sure the camera placement is right, and practice making eye contact through the camera lens.

Microphone

Getting a good picture is important. Avoiding bad audio is essential. Distracting audio undermines your presentation. Noise can make you unbearable. Noise can be in many forms: background noise (outdoor construction, nearby telephones, loud heating, ventilation, and air-conditioning (HVAC) systems), foreground noise (typing, audible breathing, lip smacking, your laptop fan), or bad audio settings (recording at too high a volume so the voice is clipped or overblown). You want a mic that accentuates your voice and minimizes noise.

Many internet searches for "best podcast microphone" or "best teleconferencing microphone" may point you to a good large-diaphragm USB condenser mic. Before immediately selecting one of those, it is important to know that there may be a far more appropriate option for you and your setup. For simplicity, the two most common types of microphones for teleconferencing are "dynamic" and "condenser" microphones. Only after determining the microphone technology should you consider the various form factors for a microphone.

Let us quickly talk about how microphones work. Sound waves move the components that convert sound to electrical signals. Dynamic microphones are typically not powered, so it is the force of the sound waves that activates the microphone. Because dynamic mics respond to stronger audio waves, they are generally better at picking up sound in the foreground and rejecting sound in the background. Condenser microphones are always powered microphones and therefore do not rely on sound waves to activate the components. People often find that similar quality condenser microphones offer warmer and fuller sound quality to their dynamic microphone counterparts. This is why condenser mics are recommended by all those internet searches. But this presupposes that you have a perfect studio. Because these mics can be sensitive, they are more prone to picking up all of those additional noises that you do not want. If you can control your environment, condenser mics are fantastic. If you cannot, dynamic mics are likely a better option for you. Do not think of dynamic microphones as a lesser microphone. They are cheaper to make, but that does not mean they are cheap. There is a very good chance that the last time you saw a singer at a concert, he or she was singing into a dynamic microphone.

A second consideration is the microphone's pickup pattern. All microphones will note the directional pattern from which they will record sound. Cardioid patterns record primarily from the front and reject sounds from behind. Omnidirectional microphones are set to record from every direction. Make sure you have a cardioid microphone for your studio setup. There are uses for other patterns, but for a setup involving a single speaker on camera, cardioid is the way to go. Some microphones allow you to switch between different pickup patterns—make sure cardioid is selected. Note that because they are more sensitive, condenser mics will likely pick up more of the peripheral sounds than the dynamic microphone with the same pattern.

The exception to a cardioid directional microphone is when you want more than one person audible. These setups, like using an omnidirectional conference room microphone, are rarely optimal and should likely be avoided. For almost every hearing, you will want one person per camera. It follows naturally to have one microphone per person. If you want your team to share a physical space, it is still recommended to have each person with individual setups. In this instance, only one microphone can be active at a time or bad feedback could occur.

The last microphone consideration is the form factor. Dynamic microphones are almost always made to talk into the end (again, think of a singer at a mic stand). Condenser microphones can be made to talk into the end or the front. Know where to talk into your microphone. Beyond a microphone on a stand, there are other forms that may better suit your needs.

Gooseneck Tabletop Microphones

These are common in courtrooms. They often have physical volume and/or mute buttons. They often come with an LED light indicator if the microphone is muted or not that doubles as indication that the microphone is powered on. These features allow you to avoid unmuting yourself in the online platform (by ensuring that the video conference is the active program and clicking the unmute "button" or, on some platforms, hitting the space bar). Instead, you can unmute on the microphone itself.

Lavalier Microphones

Lavalier microphones (a.k.a. lapel microphones) can clip on your clothing and not take any desk space. Lavaliers can be dynamic or condenser microphones. They can be wired or wireless. While some people enjoy the freedom they can provide, there are potential drawbacks to consider. First, they are so small and light it is easy to forget they are there. If you do not make sure to turn off the sound, it may broadcast some unwanted dialogue or sounds—particularly during recesses when you rush to accomplish a lot in a short amount of time. Also, if it is a wireless version, you must constantly monitor its battery. Batteries could run out of power in the middle of an examination. For this reason alone, it is often recommended to avoid anything that is battery powered—particularly for proceedings that exceed two to four hours.

XLR Microphones

These are not really a form factor, per se, but relate to the microphone's connection type. Most of the microphones recommended for video conferencing and podcasting connect directly to the computer with a USB connection. However, most standard microphones use a three-pronged external line return (XLR) connection. Off-the-shelf microphone cables have XLR ends. If you have a standard microphone for noncomputer purposes, it is likely to be an XLR microphone. These can be great value microphones or very expensive top-of-the-line microphones. While you can use XLR microphones with your computer input with good results, you need to have some way to convert the output to a signal that the computer can use. Most typically, you would convert by using an XLR to 3.5 millimeter cable or by adding a simple audio amplifier or mixer to the setup. These are fine if you are comfortable using them, but most people will opt for the ease of using one of the many USB microphone options.

Conference Microphones

These microphones can work for rooms with several attorneys who need to speak collectively without the use of headphones, and also need to avoid feedback. When this situation cannot be avoided, conference microphones are good solutions. It is critical to test the range of the conference microphone in advance.

No matter the microphone, its placement should be close to the speaker for good quality sound and low noise. It should be pointed away from any speaker output (speakers) to avoid feedback. It can be set on a table stand or mounted on a boom arm. Mics can be further enhanced with a pop filter and/or a shock mount. While it should not block your face, it absolutely can be visible on camera. At times, off-camera placement of the microphone is too far away for the best audio. You may feel like seeing the microphone is ugly or obtrusive, but it is not. Better audio is always worth the trade. In no time, a visible microphone blends into the background and nobody notices it.

Lighting

The popular webcams and microphones are often good and safe choices. Unfortunately, lighting is different. Office space with fluorescent lighting looks bad (particularly if the camera picks up the flicker). Natural sun through a well-placed window can look great, but it is unreliable: weather can fluctuate making it occasionally too bright and harsh. And as the day goes on, the shadows can shift. Even worse, the sun has a tendency to go down completely each day (more of an issue in the winter or when the proceedings happen in a distant time zone). The popular ring lights are often poor solutions, because the ring can reflect strangely in your eyes (or far worse, if you wear glasses). If your lighting has adjustable color "temperatures," the neutral or "warm" settings (3200 Kelvin (K) and below) may look more natural than "cool" temperature settings (5000 K and above).

Traditional studio lights can be bulky and can get hot. They may look great, but they may not be ideal to implement in your setup. LED panel lights can be cheap, compact, and will generate less heat. You may still opt for all sorts of accessories, such as diffusers for softer light. While you may be tempted to save space with smaller stands, be careful not to rely on stands that are too short or too flimsy.

If You Have Only One Good Light

It is best to have good, bright, flat, warm, and even light on your face. It should be placed at a high enough angle so the reflection of your eyes or glasses is not seen in the camera. But the placement should not be so high that it casts strange shadows down your face. To avoid the light being too bright, you may opt for dimmers, diffusers, or reflecting the off a bright wall or ceiling to temper the intensity.

While One Light Can Be Fine, Adding a Second Light Is Better

Position the first "key" light off-center to light most of your face. Position the second "fill" light to illuminate from the other side to counter the shadows cast by the key light. The fill light is often a slightly lower intensity for added depth. While some depth is great, having too much shadow can be distracting and will look more like an art film and less like a studio.

While Two Lights Will Yield Great Results, It Is Best to Have a Third

Three-point lighting is the studio standard. The third "back" light is directed at your head and shoulders from behind. It creates wonderful depth between you and the background. A back light (or "rim" light) should not be confused with background lighting (which lights the background rather than the subject. When setting the rim light, make sure that the bulb is not shining into the camera, as the camera's iris will close and darken the whole picture the same way it does with a window in your background, just like sunlight through a poorly placed window. The back light is oftentimes brighter than the key light to achieve the desired look. Be careful not to be too bright and cause a "halo" effect.

While Three Lights Will Make You Look Great, There Can Be More

The most important thing on camera is you. You do not want too much attention on the background (more on that below). However, simple and appropriate backgrounds can still benefit from proper lighting. Even a throw of light on an otherwise blank background can give depth. Anything to keep you from looking like you are in a void is good. Background lights can shine from off camera or sometimes they can be understated lights visible as part of the background design (like a lamp or an LED strip under a shelf).

Special Consideration for Those Who Wear Glasses

It is possible to light yourself properly and avoid glare from your glasses. Raise the lights so the reflection is not visible into the camera. Remember to test your full typical range of motion as you talk. Look up a little bit—you may have to raise the lights even more to keep the glare out. If you cannot raise the lights, you alternatively could physically block out some of the light from the bottom to remove the offending glare. Just be careful putting anything too close to hot lights. Proper lighting does no good if the building has to be evacuated.

If you are seeing glare from the reflection of the computer screen, you have a few options to correct that as well. First, you can try moving the placement of the screen, but you do not want to move it where you will have to look away from the camera to see everyone else. Another option is to lessen the light coming from the screen—lower your screen brightness and/or change the computer background to a darker color. The best option may be to brighten the lights on your face. For the same reason that a cell phone display is harder to see outside on a sunny day, cranking up the amount of light on your face will reduce the visibility of the screen reflection in your glasses.

Monitors

While passive nonspeaking viewers may be able to manage with just one monitor to watch everything, it is strongly recommended for everyone (particularly presenting attorneys and court personnel) to use two monitors to view online proceedings. In the most simple configuration, one monitor can view the other participants and the other monitor can view the evidence full-screen. Participants who need additional items on their screens—note-taking programs, email, slide presentations, and the like—may benefit from even more than two monitors. Some people swear by *many* more than two monitors. Online trials have revealed two things: you will use all the monitor space you have available; and that monitor envy is real. Get two (or more).

Monitor size is a matter of personal preference. Generally, they should be 22 inches or larger and have a minimum resolution of 1920 × 1080. Using larger panels (40–60 inches or larger) set further away can work well, too, and will mimic some of the formal distance found in a courtroom. Using desktop computer-sized screens create a more intimate option. This option is more typical and therefore more familiar for viewers in online platforms. The setup choice is a matter of preference.

Speakers and Headphones

Up to this point, all the technology described has been to enhance your broadcast fidelity. Speakers are different. Speakers have two primary functions: first, is merely for you to hear what everyone else is saying; second, is not to interfere with the audio that you are broadcasting.

For some setups, using the laptop's built-in speakers can be adequate. Upgrading to computer speakers or using a conference room's built-in audio can be fine, too. Using headphones is a good alternative to speakers. If you opt for headphones, use a set that are comfortable to wear for a long time without any pain or ear fatigue. You can choose from in-ear, over-ear, or bone conduction options. Some people prefer monitoring audio in just one ear. While you should avoid headphones that are distractingly large, do not be overly concerned with the appearance of wired headphones. Everyone understands the online platform. The wires are hardly noticed in a moment. If rock stars perform at Super Bowl halftime shows with a wired earpiece, you can too.

Speakers can cause audible feedback noise as the microphone can pick up and amplify what is playing from the speakers. The feedback loop can be fixed by adjusting the microphone or speaker placement or speaker volume. When testing your setup, make sure to test at your full speaking voice. Often people test by talking quietly, only to project much louder when giving an argument. Nothing undermines your argument like a feedback loop. Test it to avoid it. Even if you have a strong preference for speakers over headphones, it is recommended to have a backup set of headphones ready. If feedback occurs, headphones can be an easy solution as the speakers are effectively replaced.

Wireless Bluetooth headphones have advantages, but remember that you will be reliant on its batteries, which often last only two to four hours before the need to be recharged. It is probably best not to use wireless headphones, but if you do, plan to recharge or replace throughout the day. And have a backup pair of wired headphones ready if needed. Also, if your Bluetooth headphones have a built-in microphone, make sure you have your preferred microphone selected as the input to the video platform.

Backgrounds

Since most courts prohibit virtual backgrounds, it is important to address what the camera sees behind you. It is better to be too plain than too busy. Avoid bright windows visible behind you. Avoid too much personalization—particularly awards, sports memorabilia, diplomas, art, or anything distracting. Books (especially if they are nondescript), dim lamps, plants, and shelving are better options. Doors are fine, except that doors can open and have uninvited people instantly be broadcast if the doorway is on camera (in fact, everyone watching will likely notice before you do). 2020 has had many viral videos available of people who had this problem. While it may not be exciting, you can also use a backdrop (similar to the ones used at depositions or elementary school picture days).

Whatever you choose for your background, it is often best to have some space between the background and you. Because the focus should be on you, putting the background further behind will create depth. That depth will often put the background slightly out of focus. This is good. With the background out of focus, it is easy for the viewer to maintain his or her attention on the presenter.

Be aware of what it looks like with you in front of your background. Give your head some room. You do not want those nice plants to look like they are sprouting from your ears.

Optional Accessories

Document cameras can be used for still and focused viewing of physical exhibits or marking a paper document. Again, make sure your document camera can interface with your computer as many have HDMI or other audiovisual outputs, while many computers rely solely on USB inputs.

Remote presenters can give easy control of a slide presentation. Some have spotlight or mouse effects to mimic a laser pointer that might otherwise be used in a physical courtroom.

Implementation

The technology roles when in court are largely as they were pre-COVID. The online roles can generally be separated as *viewer*, *presenter*, and *host*.

Viewers

A viewer is any participant who is not speaking or presenting. This generally applies to most of the court staff, the gallery, the parties (other than their attorneys), jurors, and non-presenting attorneys. Note that viewers must still have cameras and microphones in order to communicate with the court as the need arises.

Required setup:

- Use a stable high-speed internet connection of 1.5 megabits per second (Mbps) for both upload and download (3.0 Mbps for 1080p high-definition video).
- Have a computer with webcam and microphone (preferred over tablet or smartphone).
- Familiarize yourself with accessing the platform's webcam options ("video settings").
- Familiarize yourself with accessing the platform's speaker and microphone options ("audio settings").
- Utilize a location with good lighting facing the participant with no bright lights from the background (i.e., no windows behind you).
- Secure a location free from personal distractions, people, and pets.
- Disable audio and video distractions on the computer (i.e., system sounds, phone ringers, notifications, etc.).
- Confirm you are using the current version of the operating system and teleconferencing platform (i.e., the latest version of teleconferencing software, Windows update).

Preferred setup:

- Use dual monitors (configured on the computer and the teleconference software) to see all participants and evidence when shown.

Additional best practices:

- To limit unnecessary bandwidth, have participants in the gallery turn off video and remain on mute, unless otherwise needed (unless those in the gallery are already watching via a one-way streaming platform, like YouTube).
- Log in to proceedings early to test technology for each day.

Presenters

Presenters would include judges, witnesses, presenting counsel, and trial techs. Their recommended list includes everything from the viewer's list, and adds the following.

Required setup:

- Have a computer (not tablet), webcam, and microphone (built-in microphone is acceptable).
- For screen sharing settings, select in teleconferencing platform settings to both (1) share computer sound *and* (2) optimize for PowerPoint and video, if available.

Preferred setup:

- Present from a dedicated space.
- Deselect "optimize for video clip" for better resolution of still exhibits and graphics. Only select it when actually presenting a video for a smooth playback.
- Enable audio option to "Press and hold SPACE key to temporarily unmute yourself."
- Dedicated trial presentation software (TrialDirector, OnCue, etc.) offers advanced database management and exhibit presentation tools, specifically designed for this purpose. Alternatively, documents may be screen shared using standard office document software or other viewers (Adobe Acrobat, etc.).
- If using a remote presentation controller (a "clicker"), once in slideshow view, ensure your presentation is the "active window" by clicking once on the shared screen prior to the start.
- Know how to clear your browser of internet cookies.
- Connect to the internet with wired Ethernet instead of Wi-Fi (verify speeds on sites like speedtest.net).
- Test each feature of the online courtroom prior to trial: audio, video, lighting, viewing and presenting documents, muting, viewing participants, testing virtual rooms, and so on.

Additional best practices:

- Have a backup internet connection of 1 Mbps ("MiFi" hot spot or tether computer to your phone).
- Use a second laptop to view the presentation as the rest of the courtroom sees it and/or to use as a backup trial laptop.
- Avoid using a virtual background unless required by the court. When using a virtual background, ensure that you have adequate

lighting as well as a green screen (or a very simple neutral background) to avoid visual bleeding or fading into the background.
- Review how to "pin" participants to locations on-screen; and how to "spotlight" video.
- Select the correct PowerPoint design format based on screen size displayed (likely 16:9 for wide-screen screens).
- Use a physical USB microphone with on/off toggle for better audio quality and easy unmuting.

Hosts

Hosts would be anyone tasked with the administration of the videoconferencing platform. The duties of the host may be handled by a single technology professional serving as the technical bailiff or split among several people.

Required Setup:

- Have a computer (not tablet), webcam, and microphone (built-in microphone is acceptable).
- Upgrade the platform subscription to a version that has proper encryption and, ideally, H.323/session initiation protocol (SIP) capabilities.
- Enable use of "waiting room" to admit participants.
- Select to "mute participants upon entry."
- Deselect to "allow participants to unmute themselves."
- Deselect to "allow participants to share screen"—and override this setting to allow chosen participants during the session, if desired.
- Disable private chat.
- Perform a pretrial test with all devices and all parties (separately for the jury). Confirm admitting participants, screen sharing, and moving participants to different virtual rooms.
- Create an exhibit library where the court and jurors can access exhibits as needed.

Preferred Setup:

- Use increased stable high-speed internet connection of at least 3.0 Mbps for both upload and download.
- Establish and manage virtual breakout rooms to mimic spaces in the courthouse: "courtroom," "judge's chambers/sidebar," "jury room," "gallery," and "breakout room" for each party, and an

"on-deck circle" for upcoming witnesses. Courts may simulcast the live trial to a streaming site, like YouTube, to allow public access while eliminating concerns over audience members' audio and video, and without adding to the bandwidth demands of the trial participants.

- Have a second laptop to view the presentation as the rest of the courtroom sees it.
- Monitor the bandwidth indicator. Slow connections are indicated with yellow or red bars.
- If users are accessing with a dial-up phone they may appear on the screen twice. To avoid that, hosts should consider matching or pinning the dial-up connection to the appropriate video connection.
- Be wary of using cellular for hosting—it can run nearly 10 GB of data per day.
- The court may deliver loaner laptops and/or internet hot spots for those who have insufficient or unreliable technology.
- Start the session early for all participants to log in and test devices each day.

Additional best practices:

- Verify and test pretrial procedures for admitting disputed exhibits, displaying cross-examination exhibits, and impeachment deposition.
- Courts may consider supplying all jurors with laptops, regardless of need, to avoid distractions that might exist on personal devices.
- Courts may allow jurors (or any participant) the option to report to the courthouse, where they can be provided the necessary technology, tech support, and suitable location to participate in the virtual trial.

Inevitably, additional support will be needed to assist some of the participants. Who handles this support will depend on who needs it. The technical bailiff (see Personnel section above) is most commonly the host and controls access to the proceedings, so naturally this role can provide some support for technical difficulties. It is common for the technical bailiff to have additional personnel to assist with more time-consuming tasks (e.g., a juror with connectivity problems or a witness needing to select the correct microphone). Parties may instead rely on their own trial technology support (i.e., the "hot seat" operator).

Best Practices for Technology during Online Proceedings

Online Trials Require You to Be Online

Internet connection is the primary issue that causes disruption in online trials. Test internet speed for all parties (speedtest.net is one of many options). Ensure participants have data usage plans to accommodate 2.4 GB per hour (10 GB per trial day). These amounts could be prohibitive for metered plans. Wired internet is preferable over Wi-Fi. If using Wi-Fi, a 2.4 gigahertz (GHz) connection travels farther at lower speeds, while 5 GHz provides faster speeds at shorter range.

The Presenter Always Connects Twice

Presenters may want to connect to the trial on two separate computers. One computer is the primary presenting computer. The other is used to see how the presentation appears to all the other participants. If platform controls are obstructing part of your presentation, you can address it. If you reference an exhibit or demonstrative, but forget to share your screen, you can address it. Alternatively, if you are doing everything correctly and someone voices a problem, you can be confident that the problem is on their end and not yours.

Another benefit to connecting with a second computer is that if your primary computer has a glitch, you could quickly pivot to present from the backup computer without missing a beat. If you have ever needed to reconnect to a meeting, you know the downtime feels like forever and the paranoia that you have missed something important in that moment stays with you long after you have reconnected. Having another connection stay active eliminates those problems.

Share and Share Alike

The court will likely select the online platform being used, but not always. Identify the platform well in advance of trial and who will be administering the event. Is the technical bailiff someone from the court staff or an experienced outside technical support expert? It is possible for the tasks to be split between the two: the court staff handles the litigation administration while the technology expert handles the technology.

Similarly, the court and parties should have a platform for sharing documents securely and in real time. Some videoconferencing platforms allow for file sharing, and some come with limitations. Dedicated files-haring services may have some advantages with file size and security options (as jurors should not have immediate access to share impeachment documents and gallery participants should not have access to confidential exhibits).

Court reporting services should be identified in advance. Understand if you have a real-time (or similar) feed and how to access it.

Provide contact information for IT support for issues during the proceedings. There can be different contacts depending on the issue. The technical bailiff may even use this contact information as a replacement for its webcam so it is always visible to participants.

Parties should identify how they will communicate internally. Even if the platform chat features are available, private communication should never go through the platform—even in private breakout rooms. Find an external, private, and secure method of communicating just about anything beyond "You're on mute."

Streamlining the Presentation

With real estate on the screen at a premium, nonessential participants can turn off their video for most of the proceedings. Once verified and admitted by the technical bailiff, people in the gallery can turn off their webcams, as well as non-presenting attorneys, support staff, parties, and court staff. The video feeds that remain are typically just the judge, lead counsel, the witness, and jurors. This significantly trims the visual clutter. It also has the additional benefit of easing the bandwidth demands. Jurors and other participants should be instructed to select the setting to "hide non-video participants" if available in order to decrease the clutter caused by inactive video screens.

Customizing Your View

Different platforms allow for different customization options. Typically, adjusting your screen does not alter the view for anyone else. You do not know exactly how anyone else has their screen configured. You may like watching in gallery view. Someone else may prefer pinning someone to watch the witness full screen. Some platforms allow you arrange participants. Some allow you to spotlight only one at a time, while others allow for multiple featured participants. Sometimes, platforms have additional features when multiple monitors are used. Know what options you have and select what works for you. Courts may wish to ask jurors to choose different settings for different phases of trial (e.g., "spotlight" view for an opening statement, but "gallery" view to see both the witness and the questioning attorney).

Framing Your Shot

The camera should be at eye level. If you can see the ceiling on camera, the camera is too low (shooting up to you) and should be raised. Conversely, if you can see your lap or floor on camera, the camera must be lowered. You should be zoomed into a "head-and-shoulders" shot. Your head should be prominently on camera. You want to leave some space above your head ("headroom") so it does not look like your head is bumping into the top of your monitor. Likewise, you want to visibly include shoulders in your shot. Not including the shoulders looks a lot like a floating (or decapitated) head. Including too much of your torso will make you seem too distant and you will lose some of the perceived intimacy with your audience. It is similar to a news anchor or late-night talk show host. The shot framing may be tighter when compared to many typical video-recorded depositions in order to enhance the rapport and intimacy of the presenters. A deposition's purpose is to discover evidence. A trial's purpose is to present the evidence.

What You See Is What You Get

If it is not on camera, it does not exist. This can be both an asset and a liability. A great looking studio only needs to look good from the perspective of the camera. If there is a void or a mess just off camera, it does not matter as it will not be seen. It is similar to the notion of news anchors wearing a suit top and informal bottoms—but this is a practice we do not recommend. Also remember not to reference or interact with anything that is not on camera. If the participants cannot see what you are looking at, it breaks the connection that you have worked so meticulously to create. For that moment, your elaborate setup and your polished presentation with great eye contact goes right out the window.

Testing, Testing

Similar to the need to practice to perform well, there is a need to test all the programs with all of the equipment with all of the participants. Testing does not need to happen with everyone all together. But when you test, it should be with the actual computer and accessories in the actual space and on the actual platform to be used in trial. Test days in advance with enough time to troubleshoot and procure needed items.

Controls and features can differ from platform to platform. Even something as ubiquitous as screen sharing will be different on different platforms. Even different subscription levels of the same platform can differ. One area in particular, using conference room Internet Protocol (IP) cameras, likely requires the video platform to have H.323/SIP functionality that is characteristic of office-installed videoconferencing systems, and is often only available on premium platform subscription levels.

Whenever possible use the platform computer-installed application, even if access directly through a browser is possible. The application will give you all the features available, while the browser version may lose some features and can perform differently depending on which browser you use. Since features are added and bugs are fixed all the time, make sure you have loaded the most recent version of your platform's application.

Become familiar with the platform's features and limitations. Different platforms allow for different options to control how you can arrange your screen. If you are considering using the platform's annotation feature, practice individually and with any witness who may use it. Some platforms have built-in options, like picture-in-picture, that may be ideal in certain situations. Sometimes these wonderful features are buried within menus and easy to miss. (For instance, using a PowerPoint as your background image and having a witness's camera resized in front of it.)

Have backups ready for all the essentials: a spare laptop, microphone, headphones, and webcam handy. Even a backup internet connection should be ready for essential participants—even if that is tethering to your phone (though data limits could be a problem for anything other than an emergency short-term solution).

Now That Everything Is Perfect, Lower Your Standards

Following every recommendation detailed in this chapter will create a physical space tailored for videoconferencing. But courtroom trials were not meant to be tried in equipment-filled studios. And legal professionals were not meant to be studio engineers. Each person in a courtroom finds his or her own courtroom presence. Your version of your studio should be an enhancement of your courtroom presence and not feel like a television studio in which you are trying to practice law. While talking to a camera may be uncomfortable at first, your studio should be a space in which you can become comfortable.

At this point in the evolution of online trials, it is okay if certain aspects are imperfect. What is not okay is a setup that is distracting. Inadequate audio or video is distracting. Having an unprofessional or overly busy background is distracting. Fumbling with or commenting about the technology is distracting. If you are at ease in your studio, the viewer becomes at ease watching you, and will be able to focus their attention where it belongs: on the facts and the law of your case. When you can be seen and heard properly, the technology becomes an invisible enhancement to your courtroom presence. Likewise, you become free to focus on the legal issues at hand.

Presenting online gives you an opportunity you can never have in the courtroom; you can control exactly what the viewer sees of you. Everyone sees the same visual at the same angle that is up close and personal. You have an ability to connect with participants like never before. So, fuss with your equipment a lot before the trial, and then do not fuss with it during the trial. Have support professionals tasked with running things during the session if needed. But you should relax. Relax and focus your energy on the person on the other side of the camera. Once all the gear is properly set up, the studio is meant to just fade away. Or to put it into television terms: Lights. (check) Camera. (check) Action. (that's on you)

"Look at me, everybody. I'm the first person in the world to be sent by television. Wow, what a wild trip that was. It's the greatest thing that's ever happened to me."

—Mike Teavee, *Willy Wonka and the Chocolate Factory* (1971)

Chapter 14

Managing Jurors in the Virtual Courtroom from Summons to Deliberation

Laura L. Dominic and Christopher J. Dominic

Managing jurors in an online setting at one time sounded like a daunting task, but as more and more courts were forced to adapt to changing times, it became clear that conducting a trial online is a feasible and effective endeavor. With the proliferation of web-based meetings, the art of managing people in an online setting has significantly progressed. Several courts across the country have successfully completed trials remotely, from summons to deliberation, and many will continue to do so even as we return to normal. This chapter shares the keys to successfully managing jurors throughout an online jury trial. With planning, practice, and preparation, nearly any court can administer remote jury trials effectively.

The Zoom and Cisco Webex meeting platforms have been the most widely used among the courts that have conducted virtual trials. While there are other web-based meeting platforms, the options and familiarity Zoom and Webex offer make either a good fit for the virtual trial. Many of the suggestions for managing jurors in this chapter refer to processes utilizing these services' features. However, the recommended best practices can be achieved with other meeting platforms as well.

The Virtual Courtroom: An Overview

What Does the Virtual Courtroom Look Like?

Breakout rooms are essential to the smooth operation of the virtual courtroom. Courtroom staff can move jurors from room to room within the same meeting to accomplish the many objectives of a trial. Most web-based meeting platforms have a "waiting room" function where participants wait to be admitted to the host's meeting. This function should be enabled in the virtual courtroom so that jurors are not entering a meeting where non-jury discussions are occurring. The meeting host can control when participants are let into the meeting and move them to the appropriate room. The virtual courtroom should make use of the following breakout rooms:

1. **Main courtroom**. The trial will take place in the main room. Jurors will be moved in and out of this room throughout jury selection and trial.
2. **Jury room**. Jurors will be placed in this breakout room each morning upon arrival and wait to be sent to the courtroom. This is also where jurors will be sent on breaks during the trial and where they will deliberate. Just like an in-person trial, this room should be treated as a sanctuary. Judges and attorneys should not be allowed to enter this room. Only the jury coordinator or courtroom deputy (see section Virtual Courtroom Staff, below) should be allowed to enter this room to announce when jurors will move into the main courtroom.
3. **Judge's chambers**. Most court business will take place in the main courtroom out of jurors' presence. Still, there may be times when it is easier to move the judge and attorneys into a breakout room for a sidebar than to move the jurors.
4. **Technical troubleshooting**. If a technical issue is taking longer than a minute or two to resolve, it may be more comfortable and less pressure on the juror to move them to a technical breakout room with the technical bailiff to resolve issues out of the presence of others.
5. **Plaintiff and defendant rooms**. Each side should have a dedicated room to meet with their team and client during breaks.

Virtual Courtroom Staff

Venues that have held successful virtual jury trials have reported that having a dedicated staff to attend to jurors throughout the trial is essential. Each staff member is given host or co-host capabilities to move and manage jurors during trial.

After talking with several court administrators and jury coordinators, our recommended staff roles include a jury coordinator, courtroom deputy, and technical bailiff.[1] Ideally, these roles would be assigned to different people, but functions may overlap if resources are limited:

---

1. Interview with David Slayton, Administrative Director of Courts for the Texas Office of Court Administration, and Kelley Thibodeau, Executive Assistant (Feb. 18,

1. **Jury coordinator**. The jury coordinator interacts with jurors from the summons to the start of voir dire. The coordinator provides jurors with the link to the technical check (see section Pretrial Technical Check, below) and the actual trial. When jurors arrive in the waiting room, the jury coordinator checks them in, renames their tile with their juror number for identification purposes (e.g., Mary Smith becomes Juror 5), and moves them to the jury waiting room until notice that the judge is ready to begin. The jury coordinator then moves the jury to the courtroom, where the control of juror movement is handed to the courtroom deputy.
2. **Courtroom deputy**. Once jury selection or trial begins, the jurors' movement is handled by the courtroom deputy, who remains present in the trial throughout the entire process. This person should be well versed in the technological aspects of the meeting platform. The courtroom deputy is responsible for moving jurors in and out of the courtroom throughout the day. In some venues, the judge has preferred to move the jurors; however, this should be up to the individual judge. It is highly recommended to have a courtroom deputy on staff, even in trials where the judge prefers to manage the jurors.
3. **Technical bailiff**. Jurors will inevitably experience technical issues during a virtual trial. For the trial to go as smoothly as possible, a technical bailiff is essential to ensure any issues are ironed out quickly. A technical bailiff is an expert in the meeting platform and how it operates on multiple devices, so they can troubleshoot any technical issues that arise during trial. Sometimes the difficulties are as simple as a juror being on mute, which is a relatively quick and easy fix. Other problems can take more time, necessitating one-on-one help to resolve them. Having a technical bailiff on standby to resolve issues relieves anxiety and frustration felt by everyone waiting while the problem is resolved. If a juror can no longer see exhibits, has trouble with sound, has connectivity issues, or experiences any other problem, the courtroom bailiff can move that juror to the technical troubleshooting room where they can receive individual technical help.

2021) [hereinafter Interview with Slayton and Thibodeau]; Interview with Judge Matthew W. Williams, King County, Washington, and Lisa Zimnisky, Esq., Bailiff/Judicial Assistant (Feb. 18, 2021) [hereinafter Interview with Williams and Zimnisky]; Interview with Jeff Humenik, Jury Administrator, U.S. District Court for the Western District of Washington (Feb. 22, 2021).

The Jury Summons

How Many Jurors to Summon?

The number of jurors to summon for a trial depends on many factors, and each court makes those decisions based on their own best practices. One consideration for a virtual trial is that courts have found that appearance rates have been consistently higher for online trials than in-person trials. For example, the jury administrators we interviewed reported that show rates for in-person trials were typically about 40 percent. Those rates have increased to 60 percent to 80 percent in virtual trials.[2] This is likely explained by the fact that virtual trials have eliminated many of the barriers to service that exist in physical trials (e.g., childcare, transportation, and the inconvenience of having to go to the courthouse).

Making First Contact with the Jurors

Summoning jurors for a virtual jury should generally follow the same process and procedure as a physical trial. The difference, of course, is when and how a juror reports for a virtual trial.

In most venues, the jury summons letter for a virtual trial is mailed to a potential juror's physical address. Some venues are implementing an online summons process, but most venues' current practice is to send a hard copy letter.

The summons letter should clarify that the prospective juror's physical presence is not needed and that they are being summoned to a virtual trial. In some of the early virtual trials conducted across the country, jurors complained that including the physical address to the courthouse was confusing. The letter should say in bold print near the top of the letter, "You will not be required to appear at the courthouse. Your jury duty will take place in an online format."

The summons should clearly spell out the next steps. Jurors have provided feedback to courts that they were confused about how and when to check in. Provide specific details such as:

- Inform jurors about how to complete the screening survey (many courts have a jury portal on the courthouse website where they can link a PDF form or online survey). If they cannot access the internet, provide them with a phone number they can call to complete the survey telephonically. Some venues include a hard copy screening survey in the summons letter to remove any

2. Interview with Slayton and Thibodeau, *supra* note 1.

potential barriers to service. Still, many courts prefer to collect the information electronically, and jurors often prefer to submit information electronically.

- Provide a due date for the screening survey.
- Inform them that they will receive an email or phone call by [specify date] with a date, time, and link to a "technical check" (if the court plans to implement this recommended step). (See section Pretrial Technical Check, below.)
- Inform them they will receive a link to report for jury duty on [specify date].
- Inform them that they will receive a new link for the trial after jury selection if selected as a juror.

Screening for the Ability to Serve Virtually

In addition to standard qualification questions for a physical trial, a summons letter for a virtual trial should screen for prospective jurors' technology capabilities. This will help the courts determine who can serve without the court's assistance and who may need an accommodation.

As mentioned above, many courts direct jurors to the court's online portal to complete this screening questionnaire. King County, Washington, reports that approximately 80 percent of the jurors respond electronically.[3] Those who cannot respond online can call a number to complete the survey over the phone, or if the court prefers to include a hard copy survey with the summons letter, jurors can complete and mail the survey back to the court.

Screening questions should include:

- Do you have a computer or tablet with a camera? (Accessing a virtual trial with a phone is not recommended, due to small screen size, short battery life, and limited ability to see and deliberate with fellow jurors.)
- Do you have access to high-speed or broadband internet access (in your home or other private location)?
- If you rely on internet access from a cellular provider (as opposed to cabled internet), do you have an unlimited data plan or a data plan that will allow for connection up to eight hours a day for [insert duration]?

3. Interview with Williams and Zimnisky, *supra* note 1.

- Do you have access to a private place in your home or office where you can arrange to use the internet without interruption from others?
- Do you have the ability to listen to the audio through headphones or earbuds?
- Does your computer, tablet, or headphones have a built-in microphone?
- Are you comfortable using videoconferencing technologies (i.e., Zoom, Skype, Teams, Webex)?
- Would you consider yourself reasonably proficient with videoconferencing technology?
- Are you able to appear virtually for jury service on the dates indicated in this letter?
- A link to your trial will be provided in an email. Please provide your preferred email address: ____________________.
- If you do not have an email address, please provide the best phone number where the court can reach you: ________________.

Accommodating Jurors Who Cannot Report Virtually

In no court across the country should access to technology be a barrier to service. According to the Federal Communications Commission, only 6 percent of the U.S. population lacks access to fixed broadband service.[4] Jurors work hard to access a virtual meeting if it is not readily available in their homes. Jurors have reported borrowing their child's computer or going to a friend's or family member's house with a strong internet connection. However, jurors should feel comfortable that they will not be excluded from serving if they do have connection capabilities at home. For example, in the U.S. District Court of Texas, the jury summons letter opens with, "The Court is currently permitting jurors to appear for jury service virtually using videoconferencing technologies. Individuals who cannot appear virtually will be provided alternative methods to serve, including spaces equipped with technology or the opportunity to appear in person."[5]

Accommodations can be made in at least two ways. First, the court can offer a physical space within the courthouse with a computer where

4. Federal Communications Commission, Eighth Broadband Progress Report (2012), https://www.fcc.gov/reports-research/reports/broadband-progress-reports/eighth-broadband-progress-report.

5. Interview with Slayton and Thibodeau, *supra* note 1.

the juror can report for jury duty virtually for jury selection or, if selected, throughout the duration of the trial.

Some districts can provide jurors with loaned devices if selected for a jury. A juror without access to technology reports to the courthouse on the day of jury selection and participates in voir dire from the jury assembly room or other room that can accommodate social distance. If they are selected for the trial, the court can provide them with a loaned device and a prepaid return box. Districts with enough devices have mailed the device before jury selection. The iPad with a cellular connection has been the device of choice because it offers the ability to attend trial without a hardline or wireless internet connection. The devices are restricted so that the only applications jurors can access are the meeting platform and the document review program (e.g., OneDrive or Dropbox).

A side note should be offered about jurors who are wary of technology. Some jurors are fully capable of participating in a virtual trial but are concerned about privacy or are uncertain about their ability to navigate an online platform. One juror reported to the court that they stripped down their computer of all personal information before logging into the trial. Other jurors have reported discomfort in their lack of technical sophistication and therefore find the prospect of attending trial online to be daunting. Courts can decide if these concerns should be a part of the screening and accommodation process.

Pretrial Technical Check

The pretrial technical check may be the most crucial step for a smooth and efficient trial. In the summons letter, tell jurors they will be notified of a technical check date and time. Assign a certain number of days one week before trial for this process. There should be enough time between the technical check and the first day of trial to resolve issues and arrange for accommodations if needed.

Why the Need?

Jurors who have participated in virtual trials that did not involve a pretrial technical check have offered feedback that a brief technical tutorial before jury selection would have been helpful and could have eliminated day-of delays and frustration (not only for the juror themselves, but for other jurors, attorneys, and the court).

Jurors come to trial with varying levels of experience, expertise, and comfort with technology. Absent a pretrial technical check, a significant portion of the check-in process on the first day of voir dire will be dedicated to troubleshooting with jurors who are having issues logging in, unmuting sound and turning on video, and selecting the optimal view settings, among the other problems jurors have in the online platform. Even jurors who have no issues have reported frustration waiting for the court staff to assist the jurors who do. Even though "the waiting game" is a well-known part of jury duty, jurors have reported less tolerance for delays in the virtual setting. In a physical trial, jurors expect to sit and wait, so they come prepared with a book, music playlist, or the latest podcast episode. There is nothing else to do in the jury room but sit and wait. However, when a juror is in their home, myriad distractions tempt their attention to be drawn elsewhere. These temptations create an urgency to end the waiting game more quickly in a virtual trial than in a physical trial.

Furthermore, many people report they are more proficient with technology than they actually are. Having a general familiarity with web-based meetings is not the same thing as being adept with features used in a virtual trial. People may be comfortable logging in for a social hour, but that is not the same as toggling between the meeting and a shared document file or knowing how to minimize the other participants' tiles so they can see the full screen when it is shared.

Timing

Jurors have reported that they appreciate being assigned to a technical check time slot to test equipment in advance of trial instead of the morning of the first day of trial.

The timing of a pretrial technical check should allow enough time to troubleshoot and solve technical problems, but close enough to trial that learnings during the technical check are not forgotten on the day of trial. Depending on the number of jurors summoned, the technical checks are best conducted during the week before the trial.

Expect issues with approximately a quarter of the jurors. In most cases, problems can be solved within a short amount of time. However, a pretrial tech check may reveal a needed accommodation, even among jurors who indicated on their summons questionnaire that they met the technical requirements to participate in a virtual trial.

Staffing and Breakout Rooms

On the day of the technical check, create at least three breakout rooms. At least two rooms should be for the check, and one for technical troubleshooting. Assign one staff member to check jurors in and move them to a technical check breakout room.

Consider separating participants into novice/not comfortable and expert/comfortable groups (based on the screening survey responses). This is a convenient way to manage the process. Put fewer people in the novice check room so that more time can be devoted to troubleshooting problems. The novices know they are in a meeting with other novices, which is less stressful. Furthermore, it does not waste the time of experts who understandably can become frustrated watching people struggle with something they use regularly.

The Technical Check Process

1. **Confirm they are on the device they will use and roughly in the place they will participate**. You want to make sure the device functions correctly and the internet connection is reliable. This is an excellent time to offer advice about lighting (light in front of them and not a large window behind them) and location so they are comfortable but visible on the screen.
2. **Confirm audio and video works**. Make sure the jurors can be seen and heard. Ensure the camera is clear (not dirty or covered) and that the security setting allows Zoom to access the camera and microphone. Also, confirm they can see and hear the host.
3. **Stop/start the video**. Confirm jurors know how to turn on and off their camera.
4. **Mute/unmute the microphone**. "You're on mute" may become the phrase of the 2020s. While most people know how to turn on and off their audio, few know that there are quick commands such as "push and hold the spacebar" (on Zoom for computers with keyboards) while speaking to unmute the microphone, then release the spacebar to resume the mute function. This action allows people to answer a question without manually unmuting and re-muting quickly.
5. **Speaker view versus gallery view**. Teach jurors how to toggle between the views and inform them of which view you would like them to use during jury selection and trial.
6. **Naming/renaming**. Inform them that the court staff will rename them each day. They should not change their name (there is a Zoom function that allows meeting hosts to disallow participants to rename themselves).
7. **Hiding non-video participants**. Teach jurors to hide non-video participants so those who are not "on camera" do not take up room on the grid with a blank tile.

8. **Test download and upload bandwidth**. Sharing video requires a strong bandwidth connection. In trials where video depositions or other video evidence will be shared, it is prudent to have a sample video ready to play during the tech check to ensure that jurors have a strong enough internet connection.
9. **Teach jurors how to toggle between the meeting platform and other applications or browser windows**. One of the most common problems people encounter when using a web-based conference application is maneuvering between that platform and other windows. During trial (and in deliberation in particular), jurors may need to open a second window to access and view an exhibit. When this happens, the meeting platform may be minimized or hidden. Teach jurors how to use the tabs on the bottom of the window to return back to the meeting.
10. **Explain that they will be moved around by the court staff**. Tell jurors they will be manually moved to the appropriate room at the appropriate times. Tell them they should not use the "leave room" or "leave meeting" buttons unless instructed to do so.
11. **Move jurors to the troubleshooting room if they are having difficulties**. A typical technical check should only take about five minutes. If a juror is experiencing significant difficulties, send them to the technical bailiff to get one-on-one help. The process should not be slowed down because of one juror.
12. **Inform them of the next steps**. Jurors have told court staff that they are confused about what happens next. Send the juror off with clear direction about what to expect and when.
13. **Exchange contact information**. Confirm that you have the jurors' correct contact information and jurors have the correct contact information for the court if something happens between the technical check and trial.

Voir Dire

Options for Managing Jurors during Jury Selection

Court staff have told us that empaneling more than about 24 jurors at a time in virtual voir dire is challenging to manage.[6] Groups of 12 to 20 are ideal. Although Zoom can display 49 tiles at once, with that many

6. Post-trial Interview with Judge Eric L. Dahlin, Multnomah County, Oregon Circuit Court (Apr. 1, 2021) [hereinafter Interview with Dahlin].

jurors on a screen, faces and reactions are difficult to see, particularly if the judge and counsel are not equipped with large or multiple screens. Depending on how many jurors will be called and how much time will be allotted for jury selection, we suggest two best practices for managing jurors during this trial phase.

1. **Option #1**. Assign an arrival date and time. For example, in King County, Washington, a jury pool of 60 jurors is divided into four groups of 15 jurors and are assigned to a 1.5-hour time slot on a given day. In certain circumstances, fifth and sixth panels are created as standby just in case they are needed. Jurors arrive in their time slot and go through the voir dire process. Tell the jurors at the end of voir dire that they will receive an email by a specified date and time letting them know if they were selected.
2. **Option #2**. Split jurors into two groups—one group waits in the waiting room while the other group is questioned. This option is ideal for trials calling a smaller number of jurors (25 or fewer) and allotting a short amount of time for voir dire (30 minutes or less). Give careful thought to the length of voir dire before choosing this option. Jurors have reported less tolerance for "the waiting game" in the virtual setting than in person. If jurors will be made to wait much more than 30 minutes before they are moved to the voir dire room, option #1 may be the better choice and will be appreciated by impatient jurors.

Checking In for Voir Dire

1. When jurors log in for voir dire, the following check-in process is used:
   - Jurors arrive in the waiting room.
   - The jury clerk admits them to the meeting and moves them to a breakout room.
   - If option #1 is chosen, the check-in process can be conducted in one room.
   - If option #2 is chosen, jurors are sent to a second check-in room.
2. In the breakout room(s):
   - Verify juror name and number.
   - Rename juror with just their number (since virtual trials in some venues are made public, juror names should not be visible on their tile).

- Confirm that their audio and video work.
- Ask them to select "hide non-video participants."
- Ask them to select gallery view.
- Ask them for a phone number if they lose connection.
- Provide them with a phone number they can call if they lose connection.
- Once everyone is checked in, provide instruction (see below).

Instructions to Jurors

Prepare a script to read to jurors before voir dire (this could be done in the main room with all jurors before moving them into smaller groups). Instructions should include some version of the following:

- You should be on a laptop or tablet. Smartphones are not preferred in this setting.
- It is suggested that your device should be plugged in rather than operate on battery power. If you are selected for the trial, your device will be using power for several hours each day.
- Your video camera needs to remain on at all times while not on breaks. The court would like to see you from at least the shoulders up. Please do not lie down during the trial.
- Please mute your audio. You will unmute only when answering questions or being addressed by the court.
- Please immediately inform the courtroom bailiff by raising your virtual or actual hand if you experience any of the following issues:
    - Your connection is weak (you receive a notification on your device, or your audio and/or video begins to freeze or cut in and out).
    - You are unable to hear or see the judge or attorneys.
- If you drop out of the meeting at any time, please:
    - Immediately log back in and inform the jury clerk that you need to be readmitted.
    - If you are unable to log back in, call ____ at #.
    - Upon re-admittance, inform the court of when you dropped off.
- The chat function has been disabled for this trial. Please only communicate with our court staff by raising your hand, or by telephone in the event that you have technical issues. Do not attempt to chat with any juror, attorney, or other participants in this trial through this platform's chat feature.

- Please conduct yourself in the same way you would in an in-person trial. The court expects the following:
  - You will be viewing your monitor throughout the trial and only move away from your viewing station during dedicated breaks.
  - You are free from distractions at your viewing station that will take your attention away from this trial.
  - You will not multitask while participating in this trial. We will have plenty of breaks once the trial begins to conduct outside business. This includes the following:
    - Going for a walk or running errands
    - Conversing with others in your presence
    - Answering or writing emails
    - Answering or making phone calls
    - Using the internet for anything other than these proceedings
    - Preparing meals
    - Caring for children or others in your residence
    - Doing household chores
    - Using the restroom
    - Just like an actual trial, we ask that you do not eat during the trial. You may have water or other appropriate beverages, but eating should take place only on breaks
  - No one else is in the room with you while you are participating in this trial. If anyone enters the room, please ask them to leave immediately.
  - You will not do internet research while this case is ongoing. The judge will provide further instructions on internet research.
- The court will be watching for distractions.[7] If at any time, the judge or attorneys are concerned that you are not giving this court your undivided attention, the judge or court staff may take the following action:
  - Pause the trial and speak to you directly.
  - Ask for a "360 look" at your viewing station, meaning you will turn your camera around to show the court what is in your surrounding area.

7. Virtual Trial Observation, Judge Eric L. Dahlin, Multnomah County, Oregon Circuit Court (Mar. 15, 2021).

- If you are selected as a juror, you may communicate with the court staff in one of the following ways:
  - Raise your hand virtually or physically.
  - Unmute yourself and notify the judge that you have an issue (This should only be if you are having connection issues or are otherwise missing a portion of the trial. Please do not interrupt the trial to ask any questions unrelated to your ability to see or hear the proceedings.).
  - Call ____ in the event that you are disconnected or frozen on the platform.
- If anyone has any concern that you will not be able to follow these instructions or conduct yourself in this manner, please let the court know now.

Move to Courtroom

When the judge notifies the jury clerk that the court is ready, the jurors are moved into the courtroom (If option #2 is selected, the jurors in only one of the breakout rooms are moved to the main courtroom, while the others wait in their breakout room until their turn is called. The jurors who are waiting should be told to remain at or near their viewing station to be ready when the court is ready).[8]

Rearrange the jurors (in platforms that have this feature). Once all jurors are in the courtroom, the judge and attorneys should manually move the jurors' tiles into numerical order. Note: The Zoom function only allows this operation when the "hide non-video participants" function is turned off. How you arrange your panel will not impact how other users see the jurors, but the best practice is that the judge and attorneys arrange numerically.[9]

Conduct voir dire (see Chapter 15).

Selecting the Jury

The clerk releases the last panel of the day (for option #1) or moves the jurors back to the breakout room (for option #2) so the attorneys can exercise their strikes.

In the option #1 scenario, jurors will be emailed notifying them if they have been selected or dismissed.

8. Virtual Trial Observation, Judge Marsha J. Pechman, U.S. District Court for the Western District of Washington (Nov. 2, 2020).

9. Virtual Trial Observation, Judge Thomas S. Zilly, U.S. District Court for the Western District of Washington (Jan. 25, 2021).

With the option #2 scenario, the courtroom bailiff moves the jurors not selected for the trial into the courtroom to be dismissed.

Seated Panel

In breakout room:

- Assign new numbers.
- Move to courtroom.
- Give them access to documents—Dropbox or other services.
- Reiterate instructions from above.
- Confirm best contact information.
- Provide them with court staff contact information.
- Give them the new link for trial (different than the link used for jury selection).

In courtroom:

- Swear them in. Ask jurors to confirm the oath one at a time (group audio is not always effective on a web-based platform).
- Judge takes over.

Trial

A separate meeting link should be created for the trial itself. Only the empaneled jurors and trial participants should have access to the meeting. Most virtual platforms allow the meeting to be scheduled on recurring days. This allows one link to be created for the entire trial rather than creating and sending a new link each trial day, which could easily create confusion over which link to use.

Arrival Time

Everyone attending the trial will join the meeting in the "main room." The jury coordinator can move people around once they arrive, but everyone arrives in the same space. For this reason, it is recommended that jurors be instructed to arrive at least 15 minutes before the trial team members and witnesses. This allows jurors to arrive and be moved to the jury room before trial participants arrive. Once all jury members are accounted for and placed in the jury room, the judge and trial team can join the courtroom and address pretrial issues out of the jurors' presence.

Suppose a trial team member wants to join the meeting earlier. In that case, they should immediately be sent to their respective breakout room so that they are not in the main room as the jurors arrive.

Daily Check-In Process

Each trial day begins with a daily check-in. Jurors click on the meeting link and enter the waiting room. The jury clerk admits them to the courtroom and moves them to the jury room. In the jury room, the jury clerk:

- Renames the jurors to their juror numbers
- Confirms that the jurors' cameras are on
- Confirms that audio and microphones are working
- Tells jurors not to leave their viewing station as the trial could begin at any moment
- Instructs jurors that they can (and should) talk amongst themselves about anything but the trial

Conduct in the Jury Room

After each juror has been checked in, they remain in the jury room until the judge notifies the jury clerk that the trial is ready to begin. Unless the jury clerk knows that the trial will begin later than anticipated, jurors should stay at their viewing stations with their cameras and sound on. If they leave their station, they risk being absent when the court is called to order.

While in the jury room, jurors should be encouraged to talk with each other just like they would in a physical jury room. The court staff and judges we talked to reported that the virtual setting hampered jury bonding. Jurors either muted themselves (both audibly and visually) during breaks or were uncomfortable and unsure about the socializing level that could occur. Jurors must get to know each other and form the same kinds of bonds in a virtual trial that they would in the physical trial. Initial casual interactions among the individuals enhance good group communication. In a virtual trial, that group formation can go missing without encouraging juror interaction.

Arranging Jurors in the Courtroom

Once jurors are moved to the main courtroom, it is recommended that the judge and bailiff (and attorneys if they desire) arrange the jurors numerically. This allows for a quick scan to ensure every juror is present. Someone should be assigned the task of frequently counting the jurors to ensure that everyone is present and accounted for. If a juror drops off, it is easy to determine which one is missing based on the number sequence gap.

It is also recommended that non-video participants be hidden. This reduces the number of tiles on the screen and makes it easier to see jurors.

Handling Exhibits

Exhibits can be managed through a document share site such as Dropbox, Box, Hightail, Google Drive, OneDrive, and the like. Multiple folders are created for sorting exhibits:

- Plaintiff exhibits
- Defense exhibits
- Admitted exhibits
- Jury folder (admitted exhibits, jury instructions, and verdict form)

Documents are shared on the screen during the trial, but the jury does not have access to the jury folder until deliberation. Suppose a witness needs to see the exhibit before it has been admitted. If the witness has access to read it, the parties can ask for it to be admitted and will move it to the "admitted exhibits" folder and then share on the screen.

During deliberation, every juror should be given access to the jury folder. In early trials, some courts only gave access to the foreperson but quickly recognized that this could result in undue control of the documents. Any juror can open a document with an all-access policy and ask to share the screen to call out specific portions of an exhibit.

The technical bailiff should be nearby during deliberation to help jurors with any technical difficulties they may have managing exhibits.

Observing Jurors throughout the Trial

Jurors should be instructed to keep their cameras on throughout the entire trial. The necessity of this is obvious. Jurors should also be instructed to keep their attention focused. While the judge, attorneys, and court staff can each lay eyes on the jurors, it is not practical or feasible for them to be distracted by observing the jury. For this reason, it may be best to assign a jury clerk to observe jurors during trial.

Look for any of the following distractions:

- Not looking at the camera
- Turning their camera off
- Obviously looking at something else (phone, television, another computer monitor, etc.)
- Apparently talking to someone in their proximity
- Sleeping or lying down

Any distractions should be addressed immediately. The judge can stop court and politely inform the juror to pay attention. Or the judge can call for a break and ask to speak with the juror in private if inquiry beyond a friendly reminder is needed.

Suppose an attorney or any other member of the trial team notices a distraction. In that case, it is best to notify the judge through a private channel. No attorney wants to be on record as the one who called out a juror.

What If You Lose a Juror during Trial?

As soon as it is noticed that a juror has dropped out of the trial, the proceedings should be stopped and/or go off the record. Ask the bailiff to call the juror to get them back into the meeting. Once the juror is readmitted, ask them what the last thing they heard was so they can be filled in on what they missed.

Because of the higher likelihood of losing jurors in a virtual trial due to technical difficulties, two alternates are recommended.

The judges and clerks we talked to stressed the importance of patience in the virtual trial. There *are* going to be problems. You *will* have to stop the trial more often to deal with distractions and technical difficulties. Understanding this reality will lessen the frustration when the inevitable occurs.[10]

Deliberation

Court staff have reported that deliberation is "the scariest part" of a virtual trial.[11] Until this point, the jury clerk, judge, and/or courtroom deputy have had complete control of juror management. They have had eyes on the jurors at all times and can see when distractions or difficulties occur. In the deliberation, the jury is entirely on their own. Some notes about the process:

- When the trial closes, the courtroom deputy moves the jury into the deliberation room and provides an orientation on how to access the documents.
- Instruct jurors not to begin talking about the case until they are on their own.
- Inform the jurors how and who to contact for questions or technical difficulties that arise during deliberation.
- Instruct jurors to stop deliberating if a fellow juror drops out of deliberations and to contact the court.

---

10. Interview with Williams and Zimnisky, *supra* note 1; Interview with Dahlin, *supra* note 6.

11. Interview with Williams and Zimnisky, *supra* note 1.

- Include a shared document, such as a Google Form or other live document, where jurors can type their questions to the judge. Once they have a question, they should inform the jury clerk, who can instruct the judge to open the shared document and address the question.
- Convert the verdict form to a fillable PDF form that the jury can complete electronically. They can access the form in the shared documents jury folder. The foreperson shares the screen and completes the form.

Conclusion

Even as our country is emerging from the pandemic, courts in many states have reported that the remote court proceedings may have a permanent place in our legal system. Whether a pretrial hearing or a full trial, it is likely that you will find yourself using the virtual platform at some point in the future. Following these best practices for managing jurors will help courts administer a smooth trial from summons to deliberation.

Chapter 15

Online Jury Selection

Jeffrey T. Frederick

Online or remote jury selection occurs in two basic settings: either all jurors participate online or in a "mixed" jury selection where some jurors participate online and some participate in person. This chapter will examine online jury selection in terms of how it is conducted, the advantages and disadvantages, and recommendations for future online jury selection.

Two components of jury selection are particularly salient in online jury selections. The first component is online juror questionnaires. Online juror questionnaires (either jury summons or supplemental juror questionnaires) address information gathered prior to voir dire and roughly correspond to their hard copy counterparts. Once jurors have completed any online juror questionnaires, attention turns to the second component, online voir dire and jury selection.

Online Juror Questionnaires

There are two types of online questionnaires: the juror summons and supplemental juror questionnaires. We will focus on online supplemental juror questionnaires. However, I will state from the outset that juror questionnaires are the gold standard for eliciting honest and candid answers from jurors.[1] The fact that they provide additional benefits (e.g., potential prescreening of jurors, savings in production, distribution, and postage costs, and reducing the time needed to conduct subsequent voir dire questioning) only increases their value.

1. For an in-depth discussion of jury selection and supplemental juror questionnaires, see TED A. DONNER & RICHARD GABRIEL, JURY SELECTION: STRATEGY & SCIENCE (2019); JEFFREY T. FREDERICK, MASTERING VOIR DIRE AND JURY SELECTION: GAIN AN EDGE IN QUESTIONING AND SELECTING YOUR JURY (2018); JEFFREY T. FREDERICK, MASTERING VOIR DIRE AND JURY SELECTION: SUPPLEMENTAL JUROR QUESTIONNAIRES (2018); and JURYWORK: SYSTEMATIC TECHNIQUES (NJP Litigation Consulting eds., 2d ed. 2020–2021 ed.).

Online Juror Summons

The online juror summons is discussed in more detail in Chapter 14. However, for our purposes here, these questionnaires generally address a limited number of topics including contact information, minimal demographics, qualifications for jury service, potential hardship issues, and potential barriers to serving on a jury trial, whether in person or virtual. These barriers concern the capabilities of potential jurors to serve as virtual jurors (e.g., having a private and quiet setting for the duration of trial and technological capabilities including appropriate device(s) and access to reliable internet service, among other considerations) and may also include the willingness and ability of potential jurors to serve on an in-person or virtual jury selection and/or trial. This is an important distinction for those jurisdictions that employ the hybrid method (virtual jury selection and in-person trial) and mixed jury selections (i.e., where voir dire is conducted in both in-person and virtual settings). In terms of the virtual environment, juror summons can be administered through written or electronic (text or email) notifications that include URL links and/or QR codes for completing the summons and through a jurisdiction's eJuror or juror management portal.

Online Supplemental Juror Questionnaires

Online supplemental juror questionnaires are designed to enhance, not replace voir dire. These questionnaires contain case-relevant areas of inquiry not covered in the juror summons. They include such topics as additional information on the jurors' backgrounds (e.g., educational training, occupation, contacts with law enforcement, media viewing preferences, and organizational membership) and experiences (e.g., experience with crime, lawsuits, or patents). Questions also explore relevant general and case-specific opinions and attitudes (e.g., general views on causes of crime and presumption of innocence, impressions of the parties, witnesses, or other entities, and potential areas of bias). As noted in Chapter 14, supplemental juror questionnaires should be administered sufficiently in advance of voir dire to be useful to both the parties and the court, including the ability to prescreen jurors before voir dire.

Benefits of Online Supplemental Juror Questionnaires

Online supplemental juror questionnaires, like their written counterparts, offer several benefits. They increase candor and honesty by having jurors answer questions in a private setting and minimizing what social scientists call the social desirability response bias that arises when questioning occurs in open court by high-status participants (i.e., judges and attorneys) in front of the media and public. They also allow for the streamlining of the voir dire questioning through targeted follow-up of the jurors' answers on the supplemental juror questionnaire. In addition, such questionnaires enable inquiry into certain areas (e.g., sensitive experiences or exposure to pretrial publicity) out of the presence of the other jurors, thus minimizing potential embarrassment and preventing contamination of the jury venire by jurors' bias-inducing revelations arising from pretrial publicity. Finally, research has shown that detailed online questionnaires that address case-relevant backgrounds and experiences of jurors, along with case-relevant beliefs and opinions, promote the detection of juror bias not uncovered in brief, more general online questionnaires.[2]

There are benefits unique to online supplemental juror questionnaires. Well-designed online questionnaires:

- Are easy to complete
- Produce less completion fatigue (checking/clicking answers and typing versus handwriting the answers)
- Guide the answer sequence of jurors by (1) requiring answers to certain questions before the questionnaire can be submitted; (2) allowing or preventing multiple answers to designated questions; and (3) employing "skip patterns" to direct jurors to appropriate questions and away from questions for which an answer would not be required
- Provide cost savings to the court in terms of postage, copying costs, and personnel time and effort
- Provide for easy and efficient distribution of completed questionnaires and/or their data (through spreadsheets) to the court and parties
- May increase diversity, allowing for easy and convenient initial interaction experiences with the jury management system
- Allow for easy tracking of respondents so that those who fail to complete the questionnaires are easily identified and follow-up protocols can be employed

---

2. Jessica M. Salerno et al., *The Impact of Minimal versus Extended Voir Dire and Judicial Rehabilitation on Mock Jurors' Decisions in Civil Cases*, 45 LAW & HUM. BEHAV. 336 (2021). There is no reason to believe that these effects do not extend to written supplemental juror questionnaires as well.

Use and Analysis of Supplemental Juror Questionnaires

Chapter 14 pointed out the need to administer supplemental juror questionnaires in advance of trial so their benefits can be fully utilized. The survey platforms generally used provide the parties and the court (1) electronic copies of the individual jurors' answers and (2) a spreadsheet containing all the jurors' answers. Well, how can this be used? First, attorneys can examine the individual responses of jurors to determine their favorability. Second, the database can be analyzed to understand the composition of the jury venire and the relative distribution of critical characteristics (e.g., distribution of critical opinions) across the pool and any predetermined order of jurors being considered. Third, if critical opinions/characteristics can be combined to form a scale of favorability, the database can be analyzed, and scores/ratings assigned to individual jurors. Fourth, the database can be exported into short summary forms as compared to lengthy printouts of individual responses of jurors. Finally, follow-up questions can be prepared based on the jurors' answers that streamline voir dire and make questioning more efficient.

The general flow of information is illustrated in Figure 15.1.

First, jurors complete the online juror questionnaires through an online survey platform (e.g., Google Forms and Microsoft Forms, among others). Second, the jurors' answers automatically populate a spreadsheet (e.g., Google Sheets and Excel, among others) forming the database from which various forms of analyses can be made. Finally, these databases can be used to produce summary forms that reflect all or the key answers or analyses for use during voir dire.

Final Note: Some judges and attorneys may have concerns that jurors would have difficulty in completing and/or be unwilling to complete online juror questionnaires. Results from a recent study conducted by the Maricopa County Civil Innovation Taskforce are encouraging.[3] Using two types of online questionnaires, a 46-question qualification and hardship questionnaire along with a 17-question supplemental juror questionnaire, jurors reported little difficulty in completing these online questionnaires (91 percent and 98 percent rating it "very easy" to complete the online qualification and hardship questionnaire and supplemental juror questionnaire, respectively). When it came to the jurors' willingness to complete such questionnaires if it would save time during jury selection, jurors expressed overwhelming support for such questionnaires, with 98 percent of jurors saying they would be willing to complete them (91

3. *See* Pamela S. Gates et al., *Virtual Juries: We Can but Should We? And If So, How?*, 47 LITIG. 1 (2021).

Figure 1
Basic Flow of Juror Information

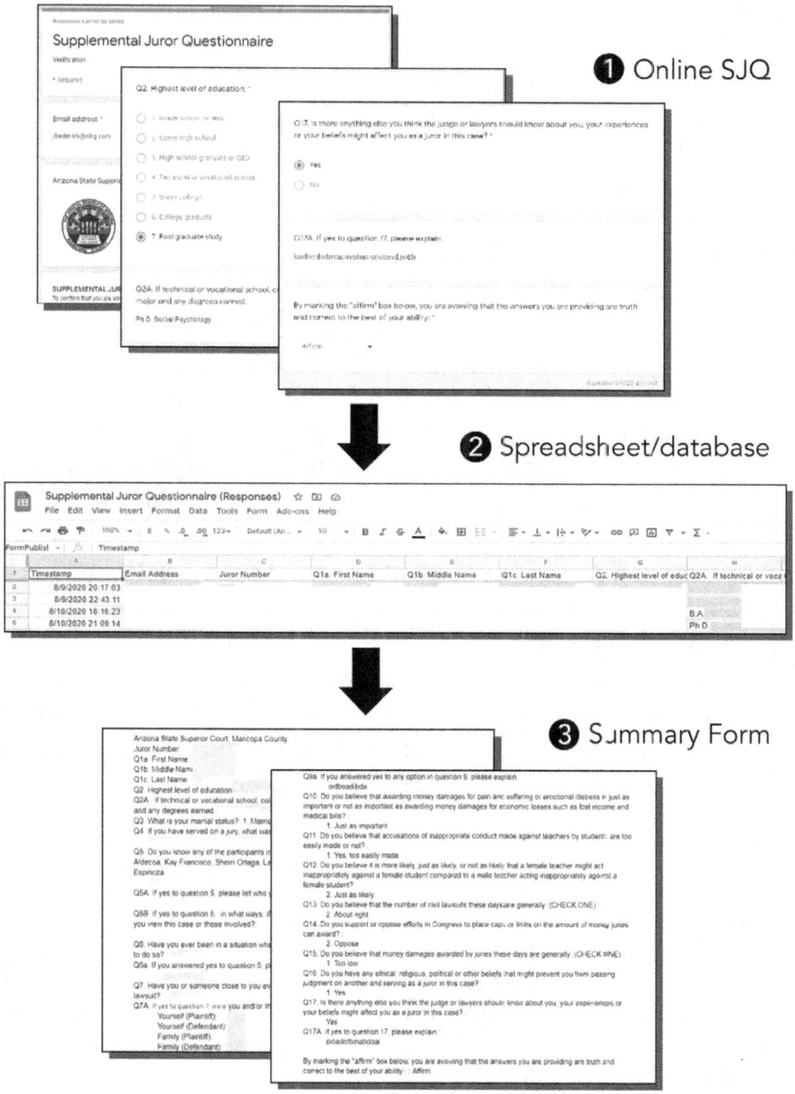

Figure 15.1

percent being "very willing"). Further evidence of the utility of online supplemental juror questionnaires comes from Judge Matthew Williams from King County Superior Court. He reports that in two recent cases, one criminal and one civil, completion rates for their respective online questionnaires were 96.3 percent and 88.5 percent in the criminal and civil trials, respectively.[4] The civil trial questionnaire can be found at the link given in Appendix F and directly at http://t.ly/wskyN. Finally, attorneys report that using online questionnaires allowed them to better prepare for voir dire, yielding a more focused and informative voir dire.[5]

To experience what it is like to complete an online supplemental juror questionnaire, I have converted the written supplemental juror questionnaire from *State v. Chauvin*[6] into an online questionnaire that is accessible through this URL: https://forms.gle/xNGcCYNY2VEa36G96.

Recommendations and Best Practices

- **Make it a joint effort**. These questionnaires provide valuable information to *both* sides. Ideally, all parties should be involved in producing the questionnaire, seeking agreement on its content. Judges are more likely to grant the use of jointly developed questionnaires.
- **Foster diversity**. Develop a strategy to include those potential jurors who might not be able or willing to complete the questionnaires online. For example, providing return mailers that include requests for written questionnaires and court phone numbers where jurors can call and request a written questionnaire will minimize potential jurors being excluded because of lack of devices/Wi-Fi or other issues.
- **Ensure data security**. While some online survey platforms (e.g., Google Forms, Microsoft Forms, and SurveyMonkey) provide "free" online surveys, make sure that appropriate steps (e.g., data encryption) are taken to ensure data security.

---

4. Personal Communication with Judge Matthew Williams, King County Superior Court (Apr. 12, 2021). According to Judge Williams, the jurors who failed to complete these questionnaires were contacted by telephone where it was determined that almost all of these missing jurors had received deferrals or failed to qualify for jury service by statute. In addition, the online questionnaire appearing at the link given in Appendix F and directly at http://t.ly/wskyN contains some case-specific questions included by agreement of the parties. Had one of the parties objected, the court would not have allowed some of them.

5. *See* Gates et al., *supra* note 3; Online Courtroom Project, Online Jury Trials: Summary and Recommendations (2020).

6. State v. Chauvin, No. 27-CR-20-12040 (Minn. 2021). The online questionnaire follows the format of the written questionnaire with some modifications for clarity in completing the process. In order to promote authenticity, I did not make significant changes to the content that would be necessary, including instructions for completing it online and the reformatting of some open-ended questions that would promote consistency with traditional online questionnaire design and efficiency.

- **Make it easy**. Provide easy access for jurors to complete online questionnaires. Take advantage of eJuror/jury management portals, URL links (using app-derived shortened versions), and scannable QR codes to quickly direct jurors to the online questionnaire.
- **Seek early completion**. Requests to complete online supplemental juror questionnaires should be sent to potential jurors weeks ahead of time, if possible, to allow for maximum completion and follow-up, along with providing parties and the courts sufficient time to adequately review and utilize the jurors' responses.
- **Keep completion windows short**. Jurors should be given short deadlines for completion of the questionnaire(s). Failure to provide short completion windows encourages procrastination and results in lower overall completion rates.
- **Monitor juror progress**. Develop a plan to monitor and reach out to jurors when their questionnaires are not completed by certain target dates. These dates should be far enough in advance of trial so that jurors have time to comply. Be prepared to follow up with potential jurors who do not complete questionnaires (or complete them incorrectly).
- **Keep text short**. Use questions in the questionnaire that are relatively short in terms of text and in the response options provided. This is particularly helpful to jurors using smartphones where display screens are relatively small.
- **Use open-ended questions sparingly**. While open-ended questions that require a narrative response are valuable in getting jurors to answer in their own words, they should be used sparingly, where possible. These questions are more difficult to answer when using devices with small screens and devices without keyboards (e.g., smartphones and tablets) and can contribute to questionnaire fatigue.
- **Do not ask leading questions**. The goal of the questionnaire is to learn about jurors, obtain information to assist in subsequent voir dire, and uncover sensitive information and potential bias. Using leading questions that signal the "correct" answer or attempt to influence jurors' original views (using push polling techniques) defeat this purpose.
- **Capitalize on platform design features**. Use mandatory or required answer functions and other questionnaire design features to facilitate complete and accurate information gathering.
- **Ensure questionnaire quality**. Thoroughly test your questionnaire for readability, functionality, and length. It is useful to

deliberately make mistakes in completing the questionnaire to see if some mistakes (e.g., skipping questions) can be prevented through the design of the questionnaire or the question format.

- **Use progress bars**. Help jurors see the "light at the end of the tunnel" by using progress bars that show jurors how far along they are in the questionnaire.
- **Provide guidance to jurors**. Develop "How to Complete Online Juror Questionnaires" videos/documents specific to your online survey platform that can be viewed online so that jurors can more easily understand how to complete the questionnaires and have answers to commonly asked questions or issues. This will help jurors who are unfamiliar with completing online questionnaires or certain online survey platforms to better navigate the online juror questionnaire, increase juror completion rates, and decrease calls to the courts regarding common survey-related questions.
- **Maximize distribution efficiency**. Develop methods for distributing the completed questionnaires or selected data/information to the parties. Online survey platforms provide spreadsheets/datasets and individual questionnaire response data in electronic documents that can be used by the court, parties, and, where appropriate, the individual jurors themselves.
- **Summarize juror information**. Parties should develop approaches (e.g., spreadsheets, exporting mail-merge data, or predesigned forms and summary sheets) to organize and prioritize the information. Also, where scales and composite scores can be generated, integrate these scores into the summary form.
- **Understand the jury pool**. Use the datasets/spreadsheets provided to analyze the jury pool to anticipate the appearance of unfavorable/undesirable jurors. Knowing what happens when a juror is removed or anticipating the future appearance of an unfavorable juror is necessary for maximizing your peremptory challenge strategy.

Online Voir Dire and Jury Selection

In general terms, the online voir dire and jury selection process is similar to the in-person process. As discussed in Chapter 14, potential jurors report to the waiting room (juror assembly room) and are called to appear in the virtual courtroom (actual courtroom) for questioning. However, the number of jurors in the virtual courtroom is limited as compared to the pre-pandemic in-person courtroom.[7] While videoconferencing platforms often limit visible display screens to 25 tiles or images, the actual number of jurors displayed on-screen is best limited to 12–15 to accommodate court and party personnel and the reduction in image size based on the number of images present.[8] Jurors are generally questioned in a group format, with the ability to address jurors individually in a separate virtual bench conference room or breakout room.

The procedures for the exercise of challenges (either cause or peremptory) are often governed by statute and/or local rules. These rules or practices did not envision a virtual environment. Because of the wide variety of practices across jurisdictions (e.g., struck versus sequential method, backstrikes, reshuffling of the venire, etc.), I will not address all variations. However, where allowed, virtual jury selection offers opportunities to select a method that works best for the parties and court.

Struck Method

Arguably the best method for jury selection, the struck method provides for the questioning of all jurors needed before any peremptory challenges are exercised. In this method, challenges for cause can be made either as questioning proceeds or at different stages during the process. The questioning of jurors ends once the total number of questioned and qualified jurors equals the number of jurors and alternates, plus the number of peremptory challenges for all parties, once any jurors successfully challenged for cause are removed.

Using the struck method, potential jurors can be questioned in panels at different times until the desired total number of jurors is achieved. Challenges for cause could be exercised at the end of each panel's questioning. Jurors who are questioned and qualified can be asked to return to the virtual courtroom at a designated time (e.g., when peremptory challenges are exercised or when further instructions are given) or may be contacted by the court and asked to return at a designated time. Peremptory challenges could be exercised in a virtual bench conference, if allowed, since jurors need not hear this and it is relatively easy to move between the virtual court and virtual bench conference.

7. As of this writing, current in-person jury selections often have limits on the number of jurors who are present in the courtroom based on social distancing requirements unless multiple courtrooms or venues such as auditoriums or other large venues are used.

8. The device used also affects the number of thumbnail images displayed. Desktops and laptops can accommodate 25, while smaller devices (e.g., tablets and smartphones) are much more limited in the number of images that can be displayed.

Sequential Method

The exercise of peremptory challenges occurs at various times before the jury is finalized in the sequential method. Courts vary widely on this practice. Some jurisdictions have one party question jurors and exercise peremptory challenges with replacements being seated and questioned before a full acceptable panel is passed to the other party. Other jurisdictions have the parties question a panel and exercise peremptory challenges at the conclusion of all questioning. The empty seats are filled with replacement jurors and the questioning and exercise of peremptory challenges continues in rounds until a final jury is seated. Still other jurisdictions use some combination or variation of the above methods.

The virtual environment's flexibility and ability to move jurors from virtual waiting rooms to the "courtroom" panel accommodates both the struck and sequential methods. However, the limitations of some platforms in terms of assigning jurors to specific locations (and maintaining that order) on the display screen can make the sequential method difficult to manage.[9]

Distinction between Mixed and All-Virtual Jury Selection Procedures

Before turning to conducting virtual voir dire, another wrinkle needs to be considered. All virtual jury selection, as the name implies, involves all potential jurors appearing via an online videoconferencing platform. The other participants (e.g., the parties, lawyers, court personnel, and judge) may appear virtually or may appear in a court environment or some combination of the two. "Mixed" jury selection refers to the situation where one or more jurors appear "in person" in the courtroom/convening facility and the remaining jurors appear virtually. Depending on how voir dire is conducted, the in-person juror(s) are either addressed simultaneously with the virtual jurors (everyone all at once), separately during the same voir dire questioning session, or separately at a different time or different day (i.e., differing reporting periods for virtual and in-person jurors). Jurors can have different levels of attention when questions are addressed globally to "panels" at different locations in the courtroom (i.e., in-court (and potentially socially distanced) jurors and virtual jurors appearing on a screen(s)). In addition, it is inherently difficult for attorneys to effectively observe all jurors at once in such settings and to pursue individual

9. It can be helpful to move passed jurors to a separate breakout or jury room. This approach would allow the parties and the court to focus their attention on unpassed jurors until the final jury is reached. Of course, jurisdictions allowing backstrikes may not find this approach workable.

follow-up questioning effectively. As such, the mixed separate questioning, different time/day approaches are most desirable as compared to the mixed simultaneous questioning approach.

Conducting Virtual Voir Dire

In general, virtual voir dire is conducted in a group setting or what might be called the *Brady Bunch* or *Hollywood Squares* setting. Many of the skills needed for effective in-person group questioning are transferable to the virtual setting. The goal of group voir dire is to have jurors participate in a candid and open manner. As such, we need to foster participation by setting the stage for the jurors (letting them know how the process will work and what is expected of them), involving jurors in the questioning process early, keeping jurors participating throughout the process, not allowing jurors to hide (e.g., avoiding participation by not volunteering answers or not raising hands when appropriate to do so), ensuring all jurors answer critical questions (e.g., not skipping questions because jurors seem reluctant to answer or are "silent"), and giving jurors an opportunity to volunteer information not previously addressed in the attorney's questions or for which a different answer is needed upon reflection. The type of questions, question wording, questioning techniques used, and attorney demeanor and style affect our ability to be successful in online voir dire (and in person, as well).

Before turning to recommendations concerning online voir dire, there are a number of differences between online and in-person voir dire settings that deserve our attention.

First, an important difference between online and in-person voir dire is that potential jurors are questioned in their own home or other non-court environment versus a courtroom. This is no small difference. Online jury selection occurs in a much less formal and more intimate environment (e.g., a bedroom or home office versus an actual courtroom). In fact, jurors have reported feeling "safer" and being more willing to be candid and participate in voir dire in the online environment.[10] This environment produces a greater *potential* for increased candor and honesty.

Second, not only can the physical environment produce greater intimacy, but the camera angle and image size can make the experience more or less intimate.[11] In general, the attorneys and jurors appear "closer" to each other given the camera angles and physical proximity to the camera.

---

10. *See* Gates et al., *supra* note 3; Michael Pressman, *Remote Jury Trials: Reporting on Judge Matthew W. Williams's Experiences in King County, Washington*, 6 Jury Matters 2 (2021); Online Courtroom Project, *supra* note 5.

11. For attorneys, close-ups appear more intimate but can be intimidating if the face is too close to the camera lens. By the same token, positioning the camera at a

Third, when the attorney focuses on the camera lens (as should be done), the jurors "see" the attorney as speaking to them since the image the juror sees is one of the attorney looking at them with associated eye contact. This can be good for "intimacy" but is bad when negative emotions or aggression by the questioner are involved. With few exceptions, it is counterproductive to argue with potential jurors no matter what the voir dire setting. It is potentially even worse to argue with a juror over an answer (e.g., the opinion is adverse to the client and the attorney feels the need to convince them otherwise) in the virtual voir dire. In essence, all jurors feel the "negative energy" directed from you since you appear to be directly "facing and speaking" to each of them.

Fourth, unlike the in-person setting, the virtual setting is a broadcast environment where lighting, visual, and sound features in the environment make a difference in how participants (both attorneys, jurors, and, for that matter, the parties) are viewed and view each other. (See Chapter 12 for tips.)

Fifth, virtual jury selection environments influence what the attorneys can see and hear during questioning. In general, the image of the juror tends to be of the torso and face along with whatever is in the background at the juror's location. While the amount of the body seen by attorneys is less than would be the case in an in-person environment, the actual image size (and degree of detail seen) is larger in the virtual environment as compared to the distance associated with the in-person environment. In addition, when safety measures are in place, such as social distancing and the wearing of masks, this further limits what attorneys can see and hear during questioning.

Sixth, beyond what is observable concerning the jurors' behaviors, the background surroundings observable by the attorney in the virtual setting can contain significant information that is absent from an in-person setting. Posters, books, pictures, artwork, and other items in the background can provide useful information concerning the jurors' values and opinions and general life circumstances. A corollary of this is that the varied backgrounds present in the virtual setting can make it more difficult to notice jurors' responses (e.g., the raising of hands) because there is no standard background behind all jurors.[12]

---

far distance can produce less intimacy and, potentially, decrease sound quality and volume. The camera placement on the part of jurors has similar effects and, additionally, affects how much of the face and body is shown in the image, which impacts the visibility of nonverbal communication cues (body language). See Chapter 12 for a more in-depth discussion.

12. The issue of backgrounds applies to attorneys (and parties) as well. As mentioned in previous chapters, consideration should be given to what jurors see of the

Seventh, while juror seating in a courtroom (even socially distanced) remains relatively fixed, the order and display of jurors in the virtual setting can change, sometimes unexpectedly. Disruptions in internet connection or device malfunction (or user error) can cause shifts in the locations of jurors. Some videoconferencing platforms do not allow the viewer to "fix" or "pin" jurors to specific locations on the display screen, making it difficult to keep track of jurors' answers during the questioning process.[13] In addition, attorneys using a "voice activated" or similar viewing option for follow-up questioning of jurors need to be aware that while the juror appears larger on-screen, there is a delay in the juror's appearance on-screen as a function of the need for the juror to make a sound and a delay in signal switch.[14]

Finally, there is an increased chance for juror distraction (e.g., use of other electronic devices, the checking of emails or texts, or conducting online searches) during questioning in a virtual setting. This potential for distraction needs to be addressed and monitored as questioning proceeds.

Recommendations and Best Practices[15]

I have designated the following recommendations to be of primary interest to the judge, attorney, or both. Of course, these participants should be aware of all the recommendations.[16]

---

attorney's surroundings. What is observable should be neutral in nature and promote, not distract from, the attorney's presentation and credibility.

13. Screen locations of jurors, unless set and fixed by the "courtroom" platform will appear differently on the screens of the attorneys and others viewing the process, which can cause confusion.

14. For example, if the juror makes a facial expression (e.g., smirks) before making a sound, the appearance of the voice-activated image is delayed until the sound is made—thus missing the facial expression. Also, if a different unmuted juror makes a sound, their image will appear, thus further disrupting this process.

15. For other discussions of best practices see Protocols for Jury Selection in Videoconference Trials 1.0. Prepared by the New York University Civil Jury Project (May 16, 2020); RIVERSIDE SUPERIOR COURT, PROTOCOL FOR REMOTE JURY TRIALS (2021); FOURTH JUDICIAL DISTRICT, REMOTE CIVIL JURY TRIAL PROJECT PILOT PROJECT (2020); Pressman, *supra* note 10; RICHARD GABRIEL, ONLINE COURTROOM PROJECT & NATIONAL INSTITUTE FOR TRIAL ADVOCACY, THE ONLINE COURTROOM AND THE FUTURE OF JURY TRIALS (2021); ONLINE COURTROOM PROJECT, *supra* note 5; Gates et al., *supra* note 3; REMOTE JURY TRIALS WORK GROUP, WASHINGTON COURTS, BEST PRACTICES IN RESPONSE TO FREQUENTLY ASKED QUESTIONS (FAQ) (2021), https://www.courts.wa.gov/newsinfo/content/Best%20Practices%20in%20Response%20to%20FAQ.PDF.

16. Some recommendations may become moot or obsolete as videoconferencing platforms evolve and become more adaptable to the courtroom setting.

Court

- **Foster diversity**. For those jurors who do not meet the requirements for online participation, the court or parties should provide what is necessary for such participation (e.g., appropriate devices, internet service, and/or kiosk, cubicle, or other physical location).
- **Develop training video/document materials**. Develop a "How to Participate in Virtual Jury Selection" video and/or document to educate jurors before they arrive (or prior to voir dire).[17]
- **Encourage larger device use**. It is difficult to position and control the camera angle with smaller smart devices, such as smartphones. Jurors should be encouraged to use tablets, laptops, or desktops, where available, to maximize the opportunity to meaningfully see jurors during questioning.
- **Use a jury assembly room**. Have jurors report to a separate virtual "waiting" room before they are moved into the virtual courtroom. Any technology issues can be addressed at this time along with the necessary attendance record keeping so that a full panel can be delivered at one time once all jurors have checked in and any issues have been resolved. In addition, allowing jurors to "chat" or otherwise interact while in the waiting room can make jurors feel more comfortable and foster eventual juror bonding.
- **Limit on-camera participants**. Turn cameras off for those who do not need an on-camera presence (e.g., paralegals, court reporter, and clerk of court). Also, have jurors set their viewing option to "active camera," if available, so critical screen space/ image size is not reduced by the display of unneeded images. This approach can help relieve bandwidth issues for participants.
- **Label jurors**. Assign an appropriate identification label to the juror's device so that jurors can be properly identified throughout the voir dire process. Either instruct the jurors how to do this or have a technical bailiff do this.
- **Have jurors use gallery view**. It is helpful to have jurors use gallery view, with key participants (e.g., judge, parties, attorneys, and court personnel) located (pinned) at the top of the gallery view with jurors appearing below them. If the online platform allows, enter jurors in order of consideration or move them to

17. See training materials developed by King County Superior Courts at King County, *Remote Participation—Video Voir Dire: How to Use Zoon Functionality*, https://kingcounty.gov/courts/superior-court/juror-information/Remote.aspx (last updated Aug. 31, 2020).

an appropriate order when jurors are "assembled" in the virtual courtroom.

- **Instruct jurors.** Jurors should be given initial instructions (with later reminders) on several topics:
  - Use the "gallery view" option (for jury selection) and do not change the view option unless directed to do so.
  - Mute the microphone unless directly answering a question or seeking to address the court.
  - Use headphones or earbuds to minimize background noise and to promote privacy.
  - Exit all non-trial related programs/tabs and do not activate any non-trial programs/platforms (e.g., email, social media, news or other internet websites) when the court is in session.
  - Disable/turn off all notifications.
  - Maintain good lighting and camera orientation as established during the tech check.
  - Procedures for reconnecting to the court if connection is lost and methods (including non-device specific channels) to notify the court if problems arise. Such procedures could be provided in advance of reporting for the tech check provided appropriate safety/privacy requirements are met to prevent unauthorized entry (e.g., Zoom bombing).
  - General instructions for jurors on what they should and should not do during questioning (e.g., do not move around, do not move camera, stay in a private setting, and do not use virtual backgrounds or filters—no kittens, please).
- **Plan for disruptions.** Disruptions in the virtual jury selection process resulting from internet connectivity issues and device failure can occur. As a result, strategies for quick resolution of disruptive events need to be developed and protocols for such events given to jurors.
- **Limit panel size.** Because videoconferencing platforms limit the number of participants who can appear on-screen (25 is a common limit), limiting the panel size to 12–15 potential jurors is advisable, plus court personnel, parties, and attorneys. While there is a trade-off between efficiency and effectiveness, it is important to remember that the fewer the number of visible tiles, the larger the visual image of the jurors will be and better able the parties are to see each juror.
- **Avoid simultaneous panel questioning.** Questioning should not occur with more than one panel at a time. Multiple simultaneous

panel questioning prevents the parties from seeing all jurors while the questions are being posed since only one panel can be displayed at a time. In addition, those jurors on the nondisplayed panels know they are not being asked for answers and, thus, their attention level is reduced. This disadvantage is not remedied by a global question to the latter panels asking if they would have answered any questions to the previous panel in the affirmative.

- **Use a "spotlight" feature**. Court and attorneys should use a "spotlight" feature, if present, to highlight the "tile" of the juror who is speaking, provided this does not disrupt or delay the video signal.
- **Schedule multiple panel sessions**. When more than one panel is needed, schedule staggered panel sessions so as to limit juror fatigue and maximize the use of the jurors' time. With jurors being at home, staggered scheduling is relatively easy and should produce maximum compliance.
- **Provide rituals**. Use activities (e.g., call court into session, remind jurors of oath, etc.) that reinforce the formality of the court process.
- **Juror break room**. Provide a virtual room, if possible, for jurors to be in during breaks or when they arrive for their appointed session. This will encourage the bonding process that typically occurs with in-person jurors. Instructions for how jurors should conduct themselves (e.g., not talk about the case) are needed along with encouragement to take an opportunity to get to know their fellow jurors.
- **Do not reveal your feelings**. Judges appear on-screen in the virtual courtroom. Jurors can and do look to see how the judge reacts during trial (and jury selection). Given the gallery view nature of the display screen, jurors are free to observe the judge without anyone noticing their focus of attention and may be influenced by any reactions exhibited by the judge.

Attorneys: Technical and Procedural

- **Know your court's platform**. Videoconferencing platforms vary widely across jurisdictions and have differing capabilities (e.g., settings, "look," and screen display characteristics). Be aware of this and make sure you have the latest update for the platform. Also, devote considerable pretrial time to practicing how you would conduct voir dire and jury selection.

- **Avoid inappropriate audio/video input**. The microphone should be muted when you are not speaking or interacting with a juror or judge. When speaking privately with team members, be sure that your audio and video feed to the virtual court has been muted/turned off.
- **Set up backchannel communications**. Develop multiple secure communication channels outside the court's videoconferencing platform along with backup plans for inevitable technical problems. Separate videoconferencing platforms or accounts, text messaging, instant messaging, email, and cellular devices should be used for private communications since the court's videoconferencing platform may not provide a truly private and secure method of communication.
- **Consider off-platform meeting "rooms**." When trial team members are in different locations, set up a virtual meeting room separate from the court for consultation. However, someone needs to monitor the "courtroom" to observe what happens when the trial team is absent to ensure that potentially inappropriate activity by jurors or parties does not go undetected.
- **Maintain good videoconferencing practices**. Jurors are viewing you during voir dire at all times. Make sure that your environment adds to, not detracts from, the jurors' impressions of you. Do not engage in behaviors that are not relevant to voir dire (e.g., answering emails, eating, walking around, etc.). Jurors will know when you are distracted and may resent your lack of engagement when they are required to sit through the process.
- **Do not forget the team**. Parties and team members who are on camera need to be aware of appropriate cautions and settings as discussed above (e.g., muting microphones, no moving around, no distracting backgrounds, no reactions to answers, etc.). As with in-person trials, jurors observe the parties at all times. It is even easier for jurors to do so in the virtual courtroom since jurors can observe everyone on camera and can do so without the other participants knowing it.
- **Capitalize on screen size**. Consider using a big-screen television, where possible, to show enlarged images of jurors while still in gallery view. Given the potential problems with "active speaker" viewing, a big screen that enlarges the jurors' images makes it easier to see their nonverbal behavior as compared to laptops and desktop computers.

- **Organize juror information**. Develop spreadsheets/charts containing relevant information on jurors along with columns to add information obtained during voir dire. If possible, add images of jurors to help you keep track of jurors' responses. This can be provided by the court, if appropriate, or generated by counsel. If allowed, screenshots of the gallery can be taken, and individual images can be placed on the spreadsheet corresponding to the jurors. The courts may consider this as an exception to the no recording rule with the explicit understanding that images will not be retained after jury selection consistent with the jurisdiction's current rules.
- **Online juror searches**. When names of jurors become available, it is important to conduct internet searches (e.g., searches through Google, social media platforms, and custom databases) of jurors, where allowed. Even if jurors' names are not made available until the voir dire questioning begins, such searches in real time may be possible and extremely valuable in providing attorneys with insights into the jurors' views, activities, and preferences.
- **Share information in real time**. Information on jurors can come from a number of sources (e.g., team notes, voir dire observations, supplemental juror questionnaires, and internet searches). Document collaboration/sharing programs like Google Docs and Microsoft Office online allow for the simultaneous editing of documents, enabling real-time sharing of information among team members. In essence, all team members can add information as it comes in and share this with other team members doing the same thing. This is particularly valuable when team members are in different locations or when using nonlocal trial consultants.
- **Identify distracted jurors**. If allowed by the court, screen snipping tools can be used to record examples of potential jurors who are distracted or engaging in inappropriate activity (e.g., viewing emails on a second display, looking down at a smartphone or the appearance of another person entering the room). These images can support assertions of inappropriate activity on the part of jurors.

Attorneys: Voir Dire Questioning

- **Use mini openings**. Mini openings provide jurors with a context for the subsequent questioning by all parties. They also serve to engage jurors in the process, leading to the potential for greater participation in subsequent questioning.

- **Have a one-sided conversation**. Capitalize on the more intimate nature of the virtual setting by using a conversational tone with jurors. This conversation should be one-sided, with jurors doing the lion's share of the speaking.
- **Familiarize jurors with the process**. There are a few process considerations different from in-person voir dire that should be made apparent to jurors. First, your camera focuses on you and if you look at your camera, you are making "eye contact" with each juror. However, if you look at each square, everyone will see the side of your face even though you feel like you are looking at a particular juror. Make jurors aware of this so they can interpret your behavior appropriately. Second, remind jurors of what to do if there are technical issues. Since you cannot see the screens of the potential jurors, ask them if they are currently having any technical issues, such as image freezing or audio delays. Third, reinforce the judge's instruction against multitasking, using other devices, or other distractions.[18]
- **Tell jurors how questioning will be conducted**. It is important that jurors understand the questioning process. Let them know how you will be asking questions. For example, you may ask general questions to the group to start a conversation on a topic and follow up with a discussion of each of their views on the topic. You will also ask more specific questions that require them to raise their hands in answer to the question. However, they need to know that, either way, you will be speaking with each of them to hear their views on the issue.
- **Raising hands versus an electronic hand raise**. Jurors physically raising their hands is preferable to the electronic hand raise. It may be difficult for some jurors to utilize the electronic hand raise efficiently. As a result, some may not be able to electronically "raise" their hands in time or decide not to participate. In addition, you cannot see the nonverbal actions associated with the raising hand (e.g., hesitation or reluctance in raising one's hand) when using an electronic hand raise function.

---

18. As mentioned earlier, it is important to identify potentially distractable jurors. If jurors engage in such inappropriate behavior and you do not reinforce this instruction or raise this issue early, the judge likely will not be receptive to any cause challenges for such jurors and think that a subsequent judicial instruction will remedy the situation.

- **Identify jurors who raise their hands**. When asking a question that involves the raising of hands, identify those who do so by reading off the names/numbers of those who raise hands and have jurors keep their hands up until their name/number is called. It is common in the virtual environment to miss a raised hand and this procedure eliminates that possibility.
- **Get jurors involved early**. While the online experience may feel more intimate to jurors, this does not mean they will all open up to you when questioning begins. You need to make sure that jurors participate early in the process. Asking each juror to answer a few basic background questions (the five-question bio: name, where they live, what they do at home or outside the home, marital status, any kids, and spare-time activities) serves to allow jurors to be more comfortable speaking in the online environment. Also, asking a question where all jurors are expected to raise their hands (e.g., a requirement of jury service such as living in the area for the required period of time) starts them participating in the process.
- **Keep them involved**. Do not rely on questions that only a few jurors will answer affirmatively. It may be great for record keeping but it reduces juror participation over time. Occasionally intersperse a majority response question. Asking a question where all jurors will raise their hands (e.g., a jury service requirement) or "reversing" a question that only a few jurors would answer affirmatively (e.g., how many of you *have not* been a victim of a crime?) can reinvigorate juror virtual participation.
- **Use springboard-and-poll method**. A valuable tool in virtual voir dire is the springboard method of questioning. In using this approach, you ask a general question to the group as a whole and follow up with any juror that answers. The answer(s) serve as a springboard for asking a different juror for their views on the issue. If no one volunteers an answer to the original question, you simply pick a juror to start the process. This process continues until no different answers surface. However, if you have not heard from every juror, you need to poll the group on which of the several viewpoints uncovered they support, thus ensuring that you know where each juror stands on the issue.
- **Do not assume jurors know who answers next**. Unless the videoconferencing platform allows the court to fix the display screen order for all jurors (which is not common as of this writing), the locations of images on the jurors' display screens will be different. As such, jurors will need to be directed to answer a question (or

asked a question by name/number) since you cannot go "down the row" when all the screens show different display orders.

- **Do not let jurors hide**. A key to good online voir dire is to make sure that all jurors participate. For those jurors who have not provided their views or participated much, direct a question to them. This will force the juror to participate and also shows all jurors that they cannot hide and avoid answering questions. You have also followed through on your original promise that you will be speaking with all jurors during your conversation.
- **Use open-ended questions judiciously**. Open-ended questions (e.g., "How do you feel about providing money compensation for pain and suffering?") are valuable in that the "right" answer is not communicated to jurors and jurors must answer in their own words. However, they can be a little tricky in the online setting. When an open-ended question is asked to a group, it can be met with silence, particularly near the beginning of questioning. Fear of "dead air" can lead attorneys to immediately shift to a closed-ended, more focused question. However, that is not necessary. If no juror volunteers a response, simply address the question directly to a juror. A valuable open-ended question should not be discarded for lack of a response since the answers, when given, can provide a more nuanced set of responses and can be followed by focused, closed-ended follow-up questions. This recommendation should not be taken to mean open-ended follow-up questions should not be used regularly. Asking open-ended follow-up questions (e.g., "Why is that?", "In what way . . .?", and "How did that make you feel?") is vital to getting a better understanding of jurors.
- **Use contrast questions**. Questions that require an agree/disagree or yes/no answer (i.e., closed-ended questions) are useful in focusing jurors' answers on specific issues. However, using questions that contrast opposing views offers additional benefits. Contrast questions include opposing views within the question and ask jurors to choose which view is closer to their own. This forces all jurors to vote on the contrast presented instead of relying on the assumption that those who fail to raise their hand "yes" really hold a "no" view.
- **Be patient and thorough**. The online voir dire environment is different and can be uncomfortable for attorneys with limited experience in remote questioning. Effective virtual voir dire requires patience and being thorough so that all key information is obtained to guide you in the exercise of challenges.

Final Thoughts

This chapter has focused on online jury selection along with the opportunities and challenges faced in this setting. It seems fitting to address the likely future of online jury selection both near-term and long-term. Online jury selection encompasses two major areas: (1) online questionnaires, in particular supplemental juror questionnaires, and (2) online voir dire and jury selection. For both of these areas, a key to future success will be the ability to bridge any digital/remote location divide as discussed in Chapter 8 by providing alternative approaches (e.g., written questionnaires were needed) and appropriate support (e.g., technology and internet connectivity support along with alternative locations such as kiosks or work areas for those jurors who need it). If these challenges can be met, online jury selection has significant potential.

Online questionnaires show both a solid near-term and long-term future. Jurors are able and willing to complete such questionnaires. Courts could see significant savings in employing online supplemental juror questionnaires as compared to written questionnaires in terms of copying, distribution, postage, and data management costs and effort, along with the ability to prescreen jurors who should not serve for various reasons (e.g., conflicts, hardships, and obvious bias). Finally, attorneys have the benefit of gaining more candid and truthful answers along with the benefits of efficient data management and, potentially, the ability to streamline and focus subsequent voir dire questioning (either in person or remote virtual). All of these factors show the promise of online questionnaires.

Online voir dire and jury selection has great potential but its future impact is more complicated whether in connection with online or hybrid jury trials. Online jury selection offers significant advantages over its health-focused in-person, masked, and socially distant counterpart. It also provides an extra layer of safety for jury trials where health concerns discourage mass gatherings (e.g., allowing the jury to be selected without needing the jury venire to report en masse). The future is more complex. The benefits of online versus in-person jury selection are (1) greater juror honesty and candor resulting from the more intimate voir dire environment; (2) the potential for greater diversity resulting from the use of remote technologies (e.g., greater attendance levels, more convenient access to the court system, and the removal of barriers arising from transportation issues, mobility issues, and the need to take off long periods of time from work or home duties);[19] (3) a greater ability to see jurors up close (visually) in their home environments; (4) more efficient jury management and scheduling (e.g., no mass gathering of jurors and related management issues, greater ease of moving jurors through the process, and greater flexibility in scheduling); and (5) greater convenience for jurors (e.g., shorter reporting times, less waiting, a perceived "safer" and more comfortable questioning environment, and no travel to and from the courthouse). A substantial majority of courts employ in-person jury trials (including in-person jury selection). It will take some time for this reliance on in-person jury selection to change. However, it is likely that more courts will recognize these benefits and adopt online jury selection as word from those courts using online jury selection becomes more widespread.[20]

19. See a summary of various judges' comments, polling, and the results of her own research in VALERIE P. HANS, VIRTUAL JURIES (Cornell Law School, Research Paper No. 21-16, 2021), http://ssrn.com/abstract= 3860165.

20. *See* Pressman, *supra* note 10; Gates et al., *supra* note 3.

Chapter 16

Online Testimony

Jeffrey Kirschenbaum

There are many reasons why online testimony can be superior to testimony in the courtroom if the online examination is done right. The trier of fact typically will pay closer attention online, allowing you to skip redundant material and focus on the important parts of your case. It will be easier for you to display key documents and emails. And you will be able to simultaneously maintain eye contact with the witness, opposing counsel, and the trial judge. This chapter will help attorneys prepare for effective direct and cross-examination at an online trial and it includes relevant information for witnesses testifying remotely.

Using the Pretrial Order to Set Ground Rules for Testimony

A pretrial order is essential for an online trial because there are some new issues that will arise that are not covered by the rules of court or code of civil procedure. Many courts and arbitration administrators have developed forms of pretrial orders for online proceedings, and trial judges have been particularly receptive to suggestions by counsel and their technologists concerning the conduct of online trials. You should begin discussions with opposing counsel about the pretrial order earlier than usual because it may affect the technology or equipment you use for the trial.

The Remote Witness Stand

To preserve the integrity of the proceeding, the pretrial order must set ground rules for the conditions under which witnesses will testify. The order should require that witnesses be seated at an empty desk or conference table, and that they not have access to any documents other than marked trial exhibits and other matters as may be approved by the court.

Witnesses should be prohibited from possessing cell phones and electronic devices or signaling to counsel or others when they are testifying.

Although a court may be reluctant to order that no one be present in the room with the witness out of concern for preserving the right to effective assistance of counsel, the best practice is for the witnesses to testify alone. The online technology works best when all participants are in different rooms and use their own computers, cameras, and microphones. You should consider including this in the pretrial order if counsel is willing to stipulate to it.

The pretrial order should prohibit the use of virtual backgrounds by witnesses and counsel, and you should not use them even if they are allowed. Virtual backgrounds are distracting and situating yourself or your witnesses in mirage is the opposite of what you want to communicate visually at trial. If counsel and witnesses are allowed to use virtual backgrounds, the pretrial order should require them to be plain and devoid of law firm advertising.

Virtual backgrounds are appropriately used by jurors. Jurors cannot be expected to rearrange their homes to eliminate visual clutter and potential distractions. Counsel should agree upon a uniform virtual background for use by jurors, and these technical arrangements should be incorporated into the pretrial order.

Trial Exhibits

There are several ways witnesses can be shown exhibits during an online trial, and it makes sense to address this in the pretrial order to avoid ad hoc decisions. Although this area is unsettled, there are good reasons to require that counsel provide witnesses with hard copies of the exhibits upon which they will be examined, to the extent practicable. It can be difficult for some witnesses to comprehend a document displayed on a screen, and it makes for a strange presentation especially when the webcam is on top of the monitor that the witness is reading from. Regardless of what the pretrial order provides, you will want your witnesses to have access to hard copies of the trial exhibits, so they are on equal footing with opposing counsel when cross-examined.

"Courtroom" Decorum

Witnesses and all trial participants should be required to dress as though they were in court and conduct themselves as though they were in court. If courtroom decorum is addressed in a typical pretrial order in the jurisdiction, it would be worthwhile to adapt these items to the pretrial order in your case.

The pretrial order may prohibit counsel from interrupting, although a court order should not be required to address this unfortunately common occurrence. Interruptions are particularly annoying online: even though two people are speaking only one can be heard.

Proper organization of the online platform is needed to preserve the formality of a trial and satisfy the litigants' right to their "day in court." During testimony, it is advisable to "pin" the trial judge, the witness, the examining attorney, and the cross-examining attorney. This arrangement will allow witness examination to proceed without the screen shifting among the participants, which can be distracting. The trial judge should be always visible, with the microphone on.

"Gallery view" can diminish the solemnity of an online trial, by reducing it to a novelty and displaying the trial judge as one participant among many. For bench trials especially, gallery view should be suppressed, although the exact solution depends upon the platform that is being used for trial.

The public has a First Amendment right to attend most trials, and this right should not be affected by using online technology. In fact, over time, online trials should lead to greater public participation in the judicial process because the courtroom can be accessed from anywhere, by anyone with an internet connection. To preserve courtroom decorum, the public should view the trial through a one-way stream such as YouTube.

Trial Objections and Sidebars

When you are in the courtroom, trial objections typically are made instantaneously, before the witness has a chance to answer the question. This is problematic online, because, if the objection is made too quickly, it will cause the questioning attorney's microphone to cut off.

The solution is for the pretrial order to require an attorney wishing to make an objection to physically raise his or her hand; and for witnesses to be directed not to answer the question if they see a raised hand. The trial judge must then recognize the attorney whose hand is raised, to inquire about and rule on the objection. This system works perfectly online and is worth considering for in-person trials.

Sidebars can be accomplished by moving the trial judge and the attorneys to a private room. The typical prohibitions on speaking objections and making offers of proof should be incorporated into the pretrial order for an online trial.

Organizing Your Trial Exhibits for Direct and Cross-Examination

It is especially important to organize your trial exhibits for an online trial. This is because pauses and delays are more irksome on video than in person. You will want to set up an electronic folder with one (and only one) copy of each trial exhibit. Store that folder where it can be accessed as quickly as possible, such as on a server or thumb drive. Do not store your trial exhibits on the cloud. When dealing with trial exhibits, you want as much computing speed as possible.

Be careful to use the exact same prefix, such as Pl. Tr. Ex. 0123, when naming the trial exhibits. If you do this, then your trial exhibits will appear in numeric order in the directory.

If your client can afford it, use an experienced trial technician to display the exhibits. If this is not possible, the exhibits can be managed by a paralegal or an attorney other than the examining attorney. Managing the video display of the trial exhibits yourself, while examining a witness, is the last resort. In order to keep your audience engaged online, you will need to deal with the exhibits with greater alacrity than if you were in a courtroom.

You will be missing out on a great opportunity for enhanced communication if you do not use robust trial presentation software, such as TrialDirector. TrialDirector pairs beautifully with Zoom. Alternatively, you can show exhibits to a witness by sharing your screen; but you probably will lack the capability to adequately enlarge and highlight the exhibits in real time with a witness on the stand.

Unfortunately, the inconvenience and expense of preparing exhibit binders remains with online trials. You will need to make at least five sets: for yourself, your remote witness stand, the trial judge, the clerk of the court, and opposing counsel. If you are conducting the direct examination of a witness who cannot testify from one of the law offices, it is advisable to deliver a set of trial exhibits to the witness or the subset of exhibits relevant to that witness's testimony. Try to avoid having a witness testify using only online documents, to avoid delays and imprecise testimony. If impeachment exhibits must be sent to an adverse witness, each one should be sealed in a separate envelope with instructions not to open unless instructed to do so during questioning.

Organizing Your Workspace for Effective Witness Examination

You will want to spend some time organizing your office for an online trial. The main objectives are to be seen and heard; and you want to maintain eye contact with the camera to the greatest extent possible. A good first step is to buy a high-definition webcam and play around with the setup. You should do this several weeks before the start of trial, to give yourself time to buy accessories, such as an external microphone, tripod, or videoconference lighting. Put the camera at eye level and bring it close enough (or adjust the controls) so your face can be seen clearly.

During the examination of witnesses, one screen will be used for the online meeting. So, your second screen can be used for viewing trial exhibits. When preparing to conduct an examination, go to your trial exhibits folder; then open the exhibits in the order you intend to use them and minimize them on the screen. The result will be a digital witness binder, consisting of PDF tiles of the trial exhibits in the correct order for your examination.

Consider buying a stand that will allow you to stack your monitors, instead of placing them side by side. By stacking the monitors, you will maximize your eye contact with the camera and avoid having to choose whether you are looking at the trial participants or your digital witness binder. You will lose your ability to control the witness and send the wrong signal to the trier of fact if you always are looking sideways.

Turn off Outlook and do not receive emails while court is in session. When you are on camera, it is obvious and disrespectful if you are reading emails. Go to your computer system's settings and turn off sound notifications and alerts.

You will want to set up some way to communicate with co-counsel and the rest of your trial team while court is in session. There are many ways to do this, such as text messaging and Slack. The best method depends on the number of people involved and their respective roles at the trial. Although text messaging is more common, Slack is advantageous because you will not be distracted by another device or perceived as checking your cell phone for messages unrelated to the trial.

You will need a complete set of trial exhibits for your office to effectively handle witness testimony. Organize the binders on a library cart or shelf near your desk, so they can be quickly accessed. While it is tempting to try to use electronic copies only, it is not possible to locate and open the exhibits quickly enough to defend a witness who is being cross-examined by an effective attorney. Also, you will want to be able to annotate the exhibits and bookmark the relevant pages of large documents.

Regarding the background, you have a little more leeway with your office than the room used for the remote witness stand. The trial participants understand that you are in your law office and expect to see some personalization. But you need to be careful not to send the wrong signal—there is more to be lost than gained here. Avoid distractions and anything that could be considered ostentatious or excessive subliminal messaging.

Preparing for Direct Examination

The use of online trial technology requires you to cover some new ground during witness preparation. First, you must allow your witnesses to practice using the technology. Replicate the conditions at trial as closely as possible and do not rush through witness preparation.

With adequate preparation, you may be able to shorten your direct examinations. The trier of fact will have no problem reading the emails and documents you introduce into evidence. Because of the nature of the technology, you will be able to watch the trial participants read the documents and emails. Therefore, the witness does not need to read the document aloud and neither do you. You also do not need to ask multiple witnesses about the same document, or email, just to emphasize its existence.

Direct examination always should be scripted and even more so for an online trial. Write out every question and identify every trial exhibit you intend to use in your outline. Sometimes, in the courtroom setting, some amount of improvisation on direct examination is needed to keep people awake. This is not the case for online trials. You already have the participants' attention; they want you to show your competence by bringing out the salient facts through thoughtful questions and answers.

The first time you refer to an exhibit, you must describe its contents and not just refer to it by its exhibit number, even though your trial exhibits will be pre-marked. Describing the exhibits is necessary to ensure a clear reporter's transcript and clerk's record. It also helps the trial participants follow the testimony. Move your exhibits into evidence as soon as a foundation is laid. By doing this, you emphasize the importance of the exhibits that you are sponsoring, and you will be alerted to any viable objections while you still have a chance to address them through the testimony of the witness on the stand.

Your witnesses should be familiar with the documents they will be shown during direct examination. When questioning a witness on direct examination about a document, first ask the witness to locate the exhibit in the exhibit binders. If there are multiple volumes, include the volume number in your outline and tell the witness where to find the document. Once your witness has authenticated the document, display it on the screen and draw the witness's attention to the pertinent portions using the online version, which everyone can see. Ask your technician to take the document down after your point has been made. You want your witness to be seen by the trier of fact while testifying, not hidden behind an exhibit.

The second screen for your remote witness stand should be placed above the camera, and not next to it. Many conference rooms have wall-mounted televisions that can be used for this purpose. Your goal is for the witness to be able to look at the document without breaking eye contact with the camera. It is a great idea to use a trial technician for your practice sessions. If your witnesses are adequately prepared, they already will have seen the cutouts and blowups and not need to ponder your questions.

It always is important for the trier of fact to develop empathy for your client and key witnesses. This can be particularly challenging for an online trial. Like all aspects of online direct examination, testimony intended to induce empathy needs to be carefully scripted and practiced. If there are personal aspects to the dispute, your witnesses should be allowed to appropriately express their feelings on direct exam.

Preparing for Cross-Examination

You should strive for shorter and more purposeful cross-examination at an online trial. The format tends to make trial participants more analytical and less receptive to mudslinging. No matter your personality or prior practices, you will look foolish if you raise your voice or pound the table while sitting alone in your office.

Online trial participants are not looking for drama for the sake of drama. They simply want the information they will need to decide the case. If you want to attack a witness's credibility, you should carefully plan. Be sure to watch any videotaped deposition you may want to use for impeachment at trial. A juicy answer on a transcript can lose its impact if it is presented as a ponderous segment of videotape. Consider reading impeachment testimony while displaying the transcript on the screen, instead of playing videotape.

In addition to identifying the exhibits you intend to use during cross-examination, you should have your technician prepare blowups and cutouts in advance. This will help you control the pace of the cross-examination. Try to get the witness to respond quickly to your questions based on the portion of the document that is being displayed on the screen.

When you are preparing your witnesses to be cross-examined, train them to refer to the exhibit binder each time they are questioned about an exhibit. This will slow the pace of cross-examination and give your witnesses more time to think. It also will prevent your witnesses from being tricked by considering only that portion of the exhibit opposing counsel believes is important.

Witnesses' "body language" remains important online. Although the area below the torso is not visible, crossed legs, crossed arms, foot-tapping, and fidgeting all impact the witnesses' image on the screen and are considered by the trier of fact. Online trials tend to favor the witnesses over counsel on cross-examination. The most important thing witnesses can do to protect themselves is to pause before answering; and a pause is quite natural during a videoconference.

The online medium creates a greater risk that your witness will fall into a trance during cross-examination and reflexively agree to a series of leading questions. Be sure to warn your witnesses against this phenomenon and encourage them to attempt to explain themselves when appropriate.

It also is important to warn your witnesses against becoming emotional on cross-examination, as always. If you are defense counsel and your clients may be called as adverse witnesses before your direct examination, instruct your witnesses to contain their emotions when being interrogated by opposing counsel and consider preparing a brief redirect even if you intend to recall your witnesses during the defendant's case in chief.

Expert Witnesses

Expert witness testimony should be tailored to the online format. Ideally, the preparations should begin at the inception of the engagement. As you work with your experts to develop their testimony, look for sources of information that will work well on-screen, such as photographs and videos.

At trial, after establishing your expert's qualifications, ask the expert to describe the work they did on the case and discuss the documents that they reviewed and the testimony they read. This will give you the opportunity to republish the key trial exhibits. You also can play excerpts of videotaped depositions or display the relevant parts of the deposition transcripts. "Show-and-tell" works better online than it does in the courtroom because the trier of fact does not have to keep shifting their attention from the witness stand to an exhibit binder or a video screen across the courtroom. Everything is right in front of them at an online trial.

Most expert witness work results in the preparation of demonstrative exhibits. Demonstrative exhibits are particularly important for online trials, where all visual communication is done on a screen. You will want to ensure that your expert's work product is appropriate for the medium.

Large spreadsheets should be broken into subparts that can be read on a computer screen. Likewise, the amount of information contained on a flowchart or a time line should be limited to that which can easily be read on a computer screen.

The direct examination of an expert witness should be heavily scripted, especially for an online trial. There is no benefit to spontaneity at this juncture. For cross-examination, the same rules apply as for percipient witnesses. You should emphasize to your expert to avoid answering questions about documents based on the information presented on the screen.

Many professional expert witnesses, such as appraisers, bring their file to the witness stand. If this is how your experts work in the courtroom, they should bring their file to the remote witness stand. However, to avoid trial objections, you will need to make an electronic copy of the file in advance and provide it to opposing counsel before your witness testifies, even if it was produced at the expert's deposition. The court and opposing counsel must know exactly what is accessible to a witness testifying remotely.

Conclusion

There are many benefits to presenting trial testimony online and understanding the benefits will help you get ready for effective direct and cross-examination. The trier of fact will pay closer attention to the testimony of your witnesses if you are prepared to move through the material in a deliberate and organized manner and maintain eye contact with the camera. To do this, you will need to carefully organize your trial exhibits and receive assistance on the day of trial from a technician or a paralegal who has been trained to use the equipment.

Many of the things discussed in this chapter can be difficult to implement remotely. For best results (and depending on the restrictions imposed on office uses in your locality), you should set up a virtual witness stand in a conference room near your office and have as many witnesses as possible testify from it. This will facilitate client conferences and witness preparation.

Begin preparing for the possibility of an online trial early in the case. Develop demonstrative exhibits that are intended for use on a computer screen; and use a detailed pretrial order to establish the ground rules to ensure that the litigants receive a fair and satisfying trial.

Chapter 17

Online Use of Trial Exhibits and Demonstratives

Noah Wick

With the recent innovation of remote jury trials comes an opportunity to take a deeper look into how we handle exchange of exhibits and demonstratives, and how we must adapt to visuals being displayed on monitors. In this chapter, we will focus on the new file sharing process with the court for remote trials, and how to still use physical demonstratives and strategies for best displaying your visuals in a remote environment.

For jury trials that have evolved to fully remote from voir dire through deliberations, judges have strongly encouraged counsel to share their materials with opposing counsel prior to starting trial. This avoids unnecessary objections and compromise on exhibits or redactions outside the presence of the court. The general sense from judges is that remote trials are more efficient when both parties resolve their exhibit discrepancies outside of court hours. Judges look favorably on counsel when they can pare down materials to only the essential exhibits worth arguing about. Thus, even impeachment materials are expected to be reviewed by opposing counsel within reason.

Typically, about one week prior to the first day of trial, a case folder on either file sharing platform will be created by either the court clerk, bailiff, or other court administrator. In federal court, the bailiff typically creates a plaintiff and defense exhibit folders on box.com. Within this folder, attorneys are encouraged to create a separate folder for impeachment materials. Both parties are initially given access to both plaintiff and defense folders. These folders and exhibits can be retrieved by both sides. This also means all impeachment materials can be accessed as well.

In King County Superior Court in Washington State, the folders are created by the Judicial Administration. These folders are treated as the clerk's official exhibit notebooks. While counsel can upload exhibits outside of court hours, once an exhibit has been added, any further alterations to the exhibit must be uploaded as a new exhibit number and cannot replace the previous version with the same exhibit number. Depending on the judge, they may prefer for you to notify the court when a new exhibit is uploaded or ask that you notify the court as soon as the next court session resumes.

Box.com versus ShareFile

Both file sharing programs have a selection at the bottom of every folder that allows you to be notified via email when an exhibit is uploaded to the folder. You can also select to be notified when a person with access to the folders downloads any item. Box.com allows for easy sharing of exhibits in a "view only" mode, which can alleviate fear of virtual witnesses downloading sensitive records. ShareFile allows for a view-only mode; however, the access must be approved by the administrator of the account—making it far more cumbersome and restrictive to use. Below is an example from a King County, Washington, remote jury trial using ShareFile to organize the exhibits, depositions, and demonstratives. Each side is provided a folder to upload exhibits including specific instructions for naming of the exhibits. The program also allows all parties to know when the exhibit was uploaded. When it is time for deliberations, the court will check mark which exhibits are to be shared with the jury and send them a link to view.

Handling Exhibits Remotely

Due to the limitations of the Zoom platform used in Washington State's King County Superior Court and Western District of Washington federal courts, counsel can only screen share with those participants in a current room. Showing exhibits to only a witness and judge is not possible during trial while in the active room. Moving the jurors back and forth to a breakout room can become tedious and slows the trial process down significantly. Due to this technological limitation, all witnesses are expected to have access to all exhibits and their own depositions when testifying. These exhibits can be physical, downloaded on a computer (preferably one that is not running the remote session), or a combination of both. Counsel is expected to ask the witness to review an exhibit and lay the foundation for admittance. During this time, the judge will also be reviewing the exhibit.

Impeachment exhibits can vary slightly depending on each judge's preference and experience with remote trials. Some judges require sealed envelopes be delivered to a witness ahead of time and clearly marked. When an impeachment exhibit is needed, the judge will instruct the witness to display the envelope via the camera verifying it is the correct one and to open it in front of the court. Typically, the jury is put back into their breakout room while this is taking place. If the judge finds the impeachment material to be appropriate, the jury will be allowed back in the primary virtual room. At this point, a brief laying of foundation in front of the jury will take place before the material is presented. Alternatively, other judges request that the witness review the material from a binder or download the exhibit digitally while on camera. Downloading exhibits is often achieved via the bailiff sending a specific link to the witness via email or directly through the remote session chat.

Illustrative Materials

Both Washington State superior and federal courts expect counsel to exchange their opening and closing presentation materials prior to the day the materials will be used in court so that objections can be efficiently argued and ruled upon. Some judges have given hard dates for opening materials, which has been an accommodation for the remote trials. In the past, these courts had counsel exchange presentations right before they started their opening argument. In person or remote, there is no nationwide standard practice on when exchanging these materials may occur as this is determined by each judge's preferences.

In a February 2021 jury trial, opening PowerPoint presentations were to be exchanged with opposing counsel no later than 5:00 p.m. on the Friday before. Any further alterations to the opening were limited to the rulings to objections discussed on the first day of voir dire. In one instance, the judge did not rule on opening objections until the first day of trial. After the rulings, court was recessed early (10:30 a.m.) for each side to adjust their openings before beginning court again at 1:30 p.m. With remote trials, this exchange practice was an accommodation that helped the trial process move forward and may be a useful example when in-person trials resume.

In King County Superior Court, opening and closing materials were required to be given an exhibit number. Any modified versions were required to be labeled with a new exhibit number, even if they were revised multiple times. In federal court, opening and closing materials were not required to have an exhibit number. Please note that even though the materials were given an exhibit number, this did not guarantee the materials would be available for the jurors to view during deliberations.

Displaying Exhibits

Some judges have required counsel to file notice of the particular presentation software used to present exhibits prior to the first day of trial. If Adobe Acrobat, Windows Media Player, or standard software is used to view files, the court may not require notification. All courts require that only the versions of exhibits uploaded to the file share platform be used. Often, counsel request their trial technicians to rebuild their database using only materials downloaded from the court's designated ShareFile or box.com folder.

With that in mind, the following recommendations:

- Plan to pre-admit exhibits.
- Assign exhibit numbers to everything in advance of trial. Even if foundation must be laid before admission, everyone should have all possible exhibits in advance.
- Be familiar with the court's impeachment exhibit sharing process. If possible, use an FTP or file sharing mechanism for new exhibits during the trial.

Physical Exhibits

Physical exhibits present an inherent problem in online trials. Ideally, there can be multiple versions to disperse among those who will need it (e.g., an accused device for each counsel, the court, and multiple witnesses). When one tries to display an exhibit by hand, it is nearly impossible to hold still enough for people to see. (Not to mention still enough for the webcam to find and maintain steady focus.) In this situation, document cameras (ELMO) are ideal to use to present the evidence. There are models that connect through USB and can be viewed as an alternative camera. The resolution is clear, and the focus stays intact. Intricate features can be pointed at with a pencil tip or a (properly manicured) fingertip. Some online platforms allow for document cameras and cameras to be viewed side by side or picture in picture to allow for a witness to be seen while personally pointing to aspects of the physical exhibit on the document camera.

Physical exhibits should be available for display to the jury for deliberations. It may be handled by the court or technical bailiff. It may be possible to send a device to each juror. Alternatively, send one device and document camera to a technically savvy member of the jury or foreperson for the entire jury to utilize during deliberations. In some remote trials, any or all of the jurors are invited to the courthouse to deliberate. As the situation demands, a viewing room with a document camera could be set up for a juror to deliberate remotely while operating a document camera for all the physical exhibits for the rest of the jury.

Demonstratives

While procedures for the most part stay similar (admissibility, disclosures, etc.), the display of them must be adjusted. In a physical courtroom, visuals are viewed in tandem with counsel and witnesses. Even on a ten-foot projection screen, visuals take a back seat. An image left displayed too long gets stale and often blends into the background as jurors will follow the action (much like a static image on a baseball field jumbotron—the fans ultimately watch the game).

In a remote setting, the documents and the people compete for screen real estate. While documents are presented, witnesses are often relegated to corners of the screen or not on-screen at all, depending on the platform and/or number of monitors available to others. When a document is shared, it pulls all the attention on it.

Types of Demonstratives

With advancements in technology, the more options we find ourselves with to not only simplify the complexity of the case, but also to entertain to keep the jury's attention. Here is a list of different types of visuals you can consider using:

- PowerPoint presentations
- Callouts and document summaries
- Graphs
- Expert curriculum vitae (CVs) and opinions
- Organizational charts
- Infographics
- Timelines
- Comparison/contrast graphics
- Analogies and metaphors

- Two-dimensional and three-dimensional still images
- Medical illustrations
- Causation and damages graphics
- Maps
- Interactive custom graphics
- Two-dimensional and three-dimensional animations

While there are several different ways to build a strong impression of your client's story, keep in mind that theme repetition will strengthen your case. Different people learn in different ways. Repeating the same message in different ways enhances the opportunity that it resonates with your audience. Figure 17.1 is an example from King County Superior Court's first remote jury trial in July 2020. This image is using the Zoom platform showing Coreen Wilson of Wieck Wilson, presenting closing to the jury. On another screen next to her she can view all of the juror's nonverbal communication and reactions to her presentation and pivot her presentation style as needed.

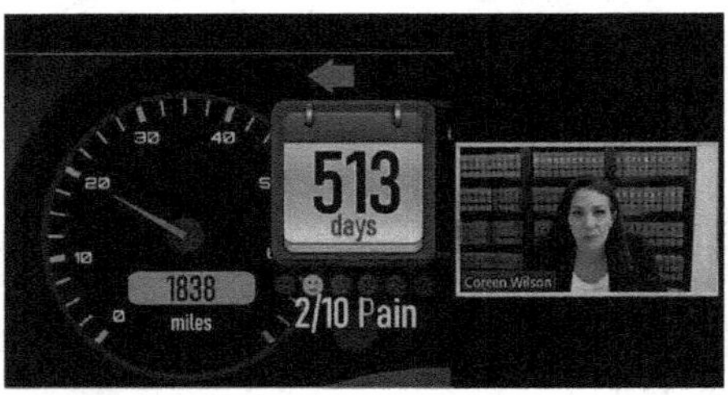

Figure 17.1

Source: Noah Wick & Coreen Wilson, Closing Argument Using Zoom in Thornton v. Warner, No. 17-2-42630-5 (Wash. Super. Ct. King County July 15, 2020).

Bigger Is Better

Similar to viewing a projection screen from a far distance, you must make sure your visuals are clear, clean, and large enough to view. In recent remote jury trials, jurors have commented that small text can be distracting from the overall impression of the visual itself. Since iPads have been one of the smallest, allowable devices to be used by jurors, we recommend you test your presentation with another account in side-by-side mode on an iPad to adjust the font size as needed. The text needs to be readable with little to no effort. Please keep in mind that while you may have perfect 20/20 vision, some of the jurors may not. While each program's font size may differ, it is recommended your font size is 14+. When creating demonstratives, do not design to the edges as windows can overlap and platform toolbars can block the edges from view. Be aware that video titles can also block part of the documents being screen shared.

Less Words, More Pictures

As they say, a picture is worth a thousand words. Instead of several bullet points, choose up to five words that summarize your strategy for each slide. Move all the words that you plan on saying into the Notes section of a

PowerPoint presentation. You can also increase the font size of these notes so you can review them more effectively as you present. Keep the jury focused on what you say and support your facts visually with showing them what you mean. Each juror has their own idea of what a word could look like if visualized. An example would be the word "dog." Each juror hearing this word has a picture of what a dog looks like but showing them a visual of the dog you are speaking about gets everyone on the same "slide." Below is an example of simplifying a common pre-procedure process of informed consent to help the jurors visualize the steps:

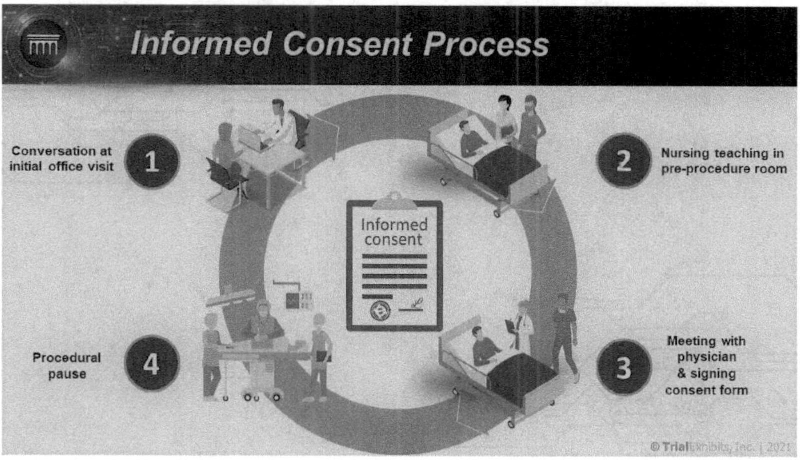

Figure 17.2
Source: Noah Wick, Trial Exhibits, Inc. (2021).

Below is an example of visually simplifying hindsight bias:

Figure 17.3
Source: Noah Wick, Trial Exhibits, Inc. (2021).

With litigation matters that involve complex medical procedures, creating visuals that help simplify the medicine allows the remote experts to build credibility with jurors during their trial testimony. Figure 17.4 is a medical demonstrative created to show the preoperative conditions of the plaintiff and the right shoulder injuries after the accident with steps of the arthroscopic repair.

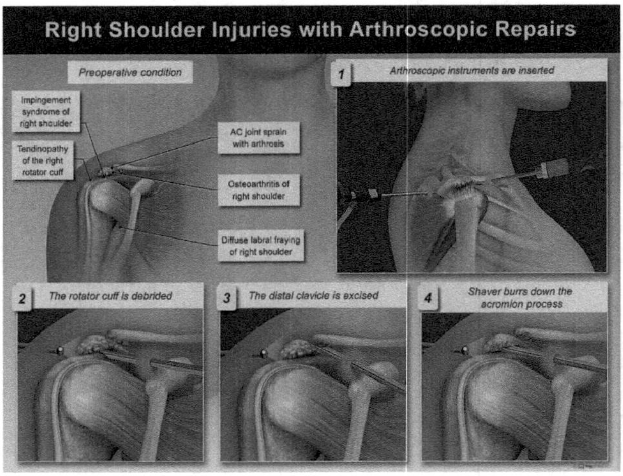

Figure 17.4

Source: Noah Wick, Trial Exhibits, Inc. (2021).

Annotations

In the courtroom, counsel often have brought oversized foam core graphic boards. During direct examination they had their witnesses and experts interact with them. In addition to a simplified visual, experts might take out a marker, draw, and teach the jury, which may then be admitted as an exhibit in some courts. This method of teaching builds significant credibility with the jury and it can be accomplished the same way using the platform's annotation feature. If the remote witness or expert needs screen control of the laptop being screen shared, there is a remote-control feature on most platforms. An example of using this would be with a physician using a Digital Imaging and Communications in Medicine (DICOM) player to explain medical imaging that involves measurements and unique tools. Keep in mind that practice makes perfect. Credibility can be lost by not understanding how to use the technology. Figure 17.5 is an example of an expert utilizing this annotation tool marking on medical imaging.

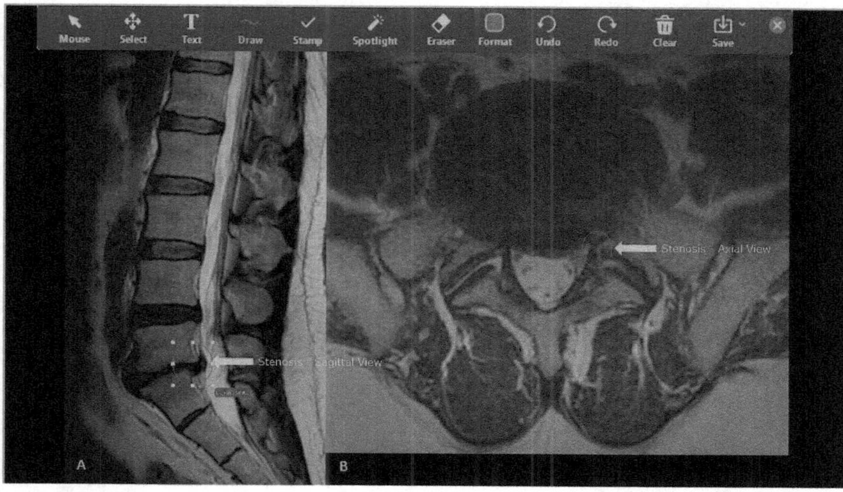

Figure 17.5

Source: Noah Wick, Trial Exhibits, Inc. (2021).

Video Playback

Platforms may have settings that trade smooth playback for image quality. When playing video, enable smooth playback mode. If playing audio, make sure your "share screen" has "share audio" enabled. If you are using a trial presentation software with synchronized text, increase the text font setting in which transcripts are displayed.

With that in mind, the following recommendations:

- Display exhibits for only the time they are needed. Be aware of when they should be removed from the screen so that focus can be put back to the witness.
- Design demonstratives to cover specific points—not bulleted lists or broad summaries intended to stay visible throughout a witness's testimony.
- You never know exactly how your shared screen is being seen on the other end, so demonstrative designs should keep font sizes as large as possible, and do not design words or text to the edges of the window screen.
- Practice with the remote platform that will be used in court—some platforms have floating toolbars near the top or bottom of the screen and directly over the title of your slide (top) or conclusion of your graphic (bottom). While lack of configuration options can be annoying, the consistency of the platform allows one to design the graphics accordingly. Use it to your advantage.

Chapter 18

Instructions for Online Juries

Justice (Ret.) Gary Hastings

Jury instructions are what inform the jury of the substantive, procedural, and evidentiary law they must apply in deciding the case. The substantive instructions let the jury know what elements must be proven with regard to the claims made, crimes charged, or defenses raised and the applicable burdens of proof attributed to the parties, if any. Procedural instructions address how jurors should approach deciding the case, which includes admonition on what they may and may not do. Evidentiary instructions address the rules relating to evidence and how the jurors should analyze the evidence, including the credibility of witnesses. Whether in person or remote, the basic jury instructions that have been used during in-person trials will remain the same and should be given. However, remote proceedings present unique issues that require additional jury instructions.

As background for this discussion, when I was on the trial court I presided over hundreds of jury trials and after each trial I invited the jurors, without attorneys, into my chambers to discuss the process and hear feedback. And, I also served on two juries while I was still active on the court of appeal. I learned in my discussions with the jurors and my actual jury service that jury instructions play a fairly minor role in the outcome of cases. More often than not jurors relied on their common sense, the facts presented, and the substantive instructions relating to the charges to come to their conclusion. Seldom did anyone discuss or mention the procedural and evidentiary instructions given.

But that is not to say they are inconsequential or unimportant. Obviously, the giving of correct jury instructions is necessary to uphold the integrity of the verdict. And on the court of appeal we did reverse a few cases when we found improper instructions were given or not given, which resulted in prejudice to the outcome. The importance of many of the instructions is to set the tone for how jurors should approach their duties. And I believe that most of them understand and take their roles seriously. I can recall only a handful of cases during my time on the court of appeal where jurors were accused of procedural improprieties such as conducting independent experiments or referring to outside information in aid of deciding the case.

An important issue related to instructing the jury is how and when instructions are given to the jury. Reading a long set of instructions takes quite awhile and can challenge jurors to maintain focus, as well as the judge. And many judges wait until the end of the trial to provide the instructions. I tried to break up the giving of instructions by instructing the jury before the beginning of trial relating to the procedural aspects of what we expect of them in how they conduct themselves and the fact they may not consider evidence from any source other than from the trial. I also gave preliminary instructions on the substantive issues involved so the jurors would have that knowledge as the evidence was produced. And if something came up during the trial for which a jury instruction was necessary, I would provide it. For example, if a deposition was used to impeach a witness I would instruct the jury about depositions. After the evidence had been received, and before closing argument, I gave further instructions to the jury relating to how they should consider the evidence and I reiterated the basic substantive instructions. After closing argument I provided closing instructions on how they should comport themselves during deliberations and how they should analyze the evidence and witnesses and reach a verdict.

By breaking up the giving of instructions I was able to impress on the jury at the beginning of trial the importance of their service and set the tone for them to approach their task and provide important information about the nature of the case they were to decide. It meant that at the end of the trial, while it was necessary to reiterate some of the substantive instructions, I did not have to read all of the instructions applicable to the case and their duties. And I provided a copy of all of the instructions to the jury so the jurors could refer to them if necessary.

As noted elsewhere in this book, remote trials can challenge the technical ability of jurors to participate. This is something that should be addressed with each juror when forming the original panel. But once the jurors have been selected, they should be instructed on the importance of their role and provided with admonitions about their conduct during trial. That would include the traditional instructions such as not discussing the case with others, not doing independent research, and waiting until the case has been finally submitted to them before they begin to form impressions or conclusions. In addition, they should be instructed that they should be in a location where they will not be interrupted and subjected to distractions, they should refrain from using other devices during the trial, and they should be shown how to communicate with the court if necessary.

In the Online Courtroom Project (OCP) demonstration trial we provided the following preliminary instructions:

> Thank you for joining us today. Jury duty is one of the most important obligations that we have as citizens in a democratic society. Because of its importance and because you will be viewing this trial online via webcam, I want to give you some special instructions in order for you to be able to best conduct yourselves as prospective jurors to ensure that the parties receive a fair and impartial hearing of the case.
>
> **Secure location and undivided attention**:
>
> First, you should secure yourself in a location where you will not be interrupted or distracted while court is in session. This will ensure that you will give your undivided attention to the proceeding. Please turn off all other electronic devices and only have the single browser window open for the link to the court proceeding while we are in session. You must be able to be seen by the parties at all times during the trial. You are also not to record any of the trial on any electronic device. If you lose a connection to the trial for whatever reason, please contact [the technical bailiff at ________] and she will give you instructions on how to log back on.
>
> **The trial process**:
>
> We will have planned breaks so please do not log off. If any of you need a break before one is announced please waive your hand to get the attention of the bailiff. We will try to be flexible.
>
> As the trial proceeds, you will hear opening statements, see witness testimony and exhibits, hear closing arguments, receive jury instructions to guide your deliberations and deliberate with your fellow jurors.
>
> In viewing the presentations of the attorneys and the witnesses, you are only to consider the evidence in the case. Do not favor one side or the other because they appear to have better computer equipment, technological proficiency, lighting, or backgrounds.
>
> If you have questions while you are listening to the witnesses, you may write them into the Q&A box on your screen. I will rule on whether the witness may answer the question. Do not hold it against either party if you do not get your question answered.

We also perceived that problems may arise during the deliberations because of the remote platform. We provided the following concluding instructions:

> **Selecting a foreperson and the process of deliberation**:
>
> Ladies and gentleman of the jury, this is the conclusion of the evidence, arguments and instructions. It is now time for you to enter into deliberations. When you do so, first elect a foreperson and then notify the bailiff of your choice. The foreperson will be sent the verdict form and you may begin your deliberations. Make sure that everyone can be heard and is able to offer their perspective. You may use the functions of virtual or visual hand raises in order to ensure that everyone is heard. You must be completely alone when you are deliberating with other jurors and must not be influenced or interfered with in any way by any outside source. It is only you twelve jurors that must decide this case.
>
> Based on the web-conferencing forum we are using, it may take extra care to manage turn-taking during your discussions so that each person is heard by all members of the jury. You may want to rely on hand-raising as well as the coordination of the foreperson in order to manage this. In the event that any of you lose your connection to the discussion, or lose the ability to clearly hear and to be heard, you should stop deliberating and await further instructions from the bailiff.
>
> **Questions for the court and the verdict form**:
>
> If you have questions or need to see specific evidence or testimony, you may communicate with the bailiff and that will be posed to the Court. In conference with the lawyers, we will decide whether the question can be answered. Do not hold it against the parties if I decide that the question cannot be answered.
>
> When you have finished filling out the form, your presiding juror must write the date and sign it at the bottom of the last page and then notify the bailiff that you are ready to present your verdict in the courtroom.

When the jury instructions are read to the jury they should be "shared" on the screen so the jury can view them as they are being read. And they should also be made available to each of the jurors for use during deliberations along with the verdict form.[1]

---

1. The Superior Court in Riverside, California, has worked up protocols for remote trials, including jury instructions. *See* RIVERSIDE SUPERIOR COURT, PROTOCOL FOR

Conclusion

We watched the jury deliberations during the OCP demonstration trial and they were similar to my experience when I served on the two in-person jury trials. It has been my experience over the years that jurors take their role seriously, try to follow the law as instructed, and usually reach a result that is supported by the law and evidence. I do not believe that holding trials remotely will challenge that perception.

Remote Jury Trials (2021), https://riverside.courts.ca.gov/GeneralInfo/AttyLitigants/Remote-Civil-Jury-Trials-Protocol.pdf.

Chapter 19

Socially Distanced and Online Jury Deliberations

Karen Lisko

Introduction

The deliberative process in a jury trial is its very cornerstone. No matter what the medium by which jurors come together to reach their verdict, the integrity of jury deliberations must be protected. The right concern is to ensure that jury deliberations proceed effectively when the "jury room" is virtual. Can jurors freely express their perspectives? Can jurors with varying technical acumen adequately see and hear their fellow jurors' comments? Can they collectively process the evidence, visually and otherwise? Many courts across the country have used sound approaches to protect the process and our own virtual jury research has illuminated consequences of the virtual deliberation medium, both positive and negative. Understandably, the deliberation phase of a remote jury trial raises questions because jurors are on their own, unmonitored, and are only connected by technology. Despite those concerns, reports from jurors and court staff in recent online trials indicate that jurors appear to have risen to the occasion.[1] This chapter addresses recommended practices for protecting the jury's deliberative process in the virtual deliberation room.

The New Challenges for Jury Deliberations during the Pandemic

Since the formation of the U.S. Constitution, the American jury system has included jury deliberations with interestingly little direction other

1. For example, Confidential Interviews with Oregon Jurors Who Participated in First Remote Jury Trial in Multnomah County, Oregon (Apr. 2–6, 2021) [hereinafter Confidential Juror Interviews].

than that described in jury instructions regarding things like decision rules, ranging from requirements of unanimity to something greater than a majority. As a result, many courts have intuitively relied on their own experience to address extralegal needs around deliberations such as ensuring jurors have ample opportunity to get to know one another before deliberations or that they have the privacy they need to deliberate confidentially and without outside influence. Our interviews with a number of judges and litigants given the disappearance of traditional (non-socially distanced) face-to-face deliberations during the pandemic have revealed deep concerns about protecting the deliberative process. Fortunately, remedies exist when courts apply best practices to socially distanced deliberations and the online deliberations.

What We Know about Jury Deliberations in the Socially Distanced Courtroom

Three sources inform our knowledge about jury deliberations in the socially distanced courtroom: (1) interviews with jurors who have deliberated in actual cases, (2) interviews with mock jurors who have deliberated in simulated cases, and (3) interviews with judges who have presided over both types of cases. Since few socially distanced jury trials occurred in 2020, the data is too small to enable us to call it statistically significant. Still, the trends are meaningful for determining best practices.

The Consequences

The group dynamic that ultimately culminates in a jury's deliberations starts well before the jury shuts it door to render its verdict. In our work and discussions with court personnel about socially distanced jurors, two immediate issues arose, one with jury bonding and one with physical logistics.

Jury Bonding. In traditional jury trials, the jurors typically begin trial as strangers and gradually get to know one another as they gather in the jury room on breaks. While not all jurors bond with one another, they are at least familiar with one another by the time they begin to deliberate. This familiarity can ease the process of discussing the case during the deliberative process. In the socially distanced trial where jurors must distance six feet apart per guidelines from the Centers for Disease Control and Prevention (CDC), it is reasonable to conclude that the typical jury deliberation room cannot accommodate all jurors. Some courts have used overflow space for a few jurors (even including the hallway outside the jury room) but it creates the exigency of "subsetting" jurors rather than allowing them to get to know one another as a single group.

Handled poorly, this reality means factions of the jury could get to know one another better than others.

Juror Candor. At the point of deliberations in the socially distanced courtroom, most courts have made the entire courtroom available to jurors so they could deliberate as a group. While this physical space technically checks the box for social distancing, one challenge some jurors face is being seen and heard in the physically spaced arena since they are also masked and often not clearly in one another's line of vision. Our observation was that it takes an extra level of assertiveness for jurors to speak up and contribute to discussions. This skill can already prove challenging for quieter jurors. The socially distanced courtroom makes it even more difficult.

Social scientist Jeffrey Frederick and I supported Judge Pamela Gates and the Superior Court of Arizona in running jury simulations that compared socially distanced jury trials using personal protective equipment (PPE) with remote jury trials.[2] When we asked jurors about the effect of social distancing and PPE on their ability to express their own views during deliberations, they reported the following:

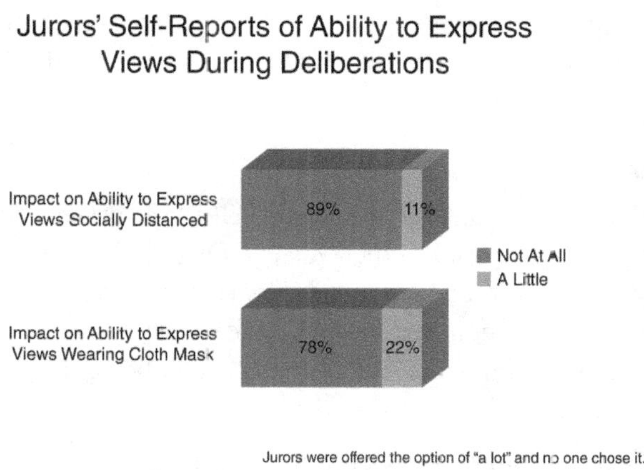

Figure 19.1

The number of jurors participating in this in-person, socially distanced simulation was small but unique. Many of these jurors had served as jurors in actual courtrooms prior to the pandemic, allowing many of them a special point of comparison. In addition, we asked these jurors how they believed social distancing and the use of PPE affected their

2. *See* Pamela S. Gates et al., *Virtual Juries: We Can but Should We? And if So, How?*, 47 Litigation 1 (2021).

peer jurors' ability to express their views during deliberations. While they generally rated their peer jurors' abilities similarly, some felt their peer jurors had a bit more trouble in comparison to themselves.[3]

The Opportunities

In conversations, a few judges have expressed the sentiment that they believe jurors who have shown up for jury duty in physical courtrooms during the pandemic have taken their jobs especially seriously. They reason that jurors appreciate the rare nature of such proceedings and that jurors' own efforts in going through extra safety precautions likely reflects a real commitment to jury duty.

What We Know about Jury Deliberations in the Virtual Jury Room

Reports of COVID-19 infections in some socially distanced courtrooms[4] despite best efforts has driven many courts to either suspend jury trials indefinitely or to conduct jury trials remotely. By the time the jury deliberation phase occurs in the remote trial, jurors have often attained some level of technical proficiency. Still, a deeper look at this medium reveals a variety of experiences both related and unrelated to technical acumen. And why wouldn't it? Remote jury deliberations were unheard of before 2020.

Jurors' Experiences

Getting to Know One Another. In February 2021, jurors deliberated in the first remote patent jury trial in the Western District of Washington. Juror David Lee had the combined, unique experience of participating in that trial and being able to contrast his remote experience with his in-person jury trial experience in 2016. He compared socializing between the two mediums this way:

> The jury was asked to connect to the Zoom call up to 30 minutes prior to resumption of the trial each day. Clerks stated at least once that this was so that jurors could get to know each other. Items behind jurors (pets, posters, guitars, etc.) provided great starting points for conversation. Socialization wasn't as good over Zoom as it was for me in person at the prior trial, but the

3. *See id.*

4. R. Robin McDonald, *Courthouses Close Anew as Staff, Judges Test Positive for COVID-19*, Law.com, July 9, 2020, https://www.law.com/2020/07/09/courthouses-close-anew-as-staff-judges-test-positive-for-covid-19/.

> duration for this trial was much shorter, so that's expected. I feel like socialization was in a good place prior to deliberations.[5]

Use of Exhibits. As we find in the great majority of jury trials, remote juries are hungry to make effective use of case exhibits. Jurists from two different jurisdictions in the state of Washington expressed feelings about their remote access to exhibits. Judge Sean O'Donnell, chief criminal judge in King County, Washington, recently presided over a Zoom jury trial. He described the jurors' desire to interact more with the electronic exhibits than they could in that trial. They wanted to be able to share exhibits from their own computers, mark them with circles, and show them side by side to one another. In other words, they wanted to be able to do more than just possess them.[6] In a Western District of Washington remote trial, juror David Lee said that jury accessed electronic exhibits via box.com, which he found preferable to his prior in-person jury experience "where all jurors must crowd around a single copy of the documents."[7]

The Consequences

The Loss of In-Person Connectivity. Juror David Lee felt less satisfied with the final moments of trial: "The only point at which it would have been dramatically better to be in the same room was after the trial so that I could say goodbye to each fellow juror. It felt strange to just collectively wave goodbye and leave the meeting."[8]

Litigators' Concern over Diminished Persuasiveness. Ricky Raven is a partner with Reed Smith LLP and was defense counsel in a nine-week-long asbestos trial held over Zoom in Alameda County, California. He recounts the process by which jurors deliberated and his concerns regarding their choice of forum:

> The jurors were split 50/50 on deliberating at home separately or collectively at the courthouse. The court decided to allow them to deliberate separately, to which I was opposed. However, they all thought deliberating from home was better because they could control their individual environments. Also, I think juror

5. Cara Salvatore, *Clear Exhibits, Abrupt Goodbyes: Life As a Zoom Juror*, Law360, Feb. 9, 2021, https://www.law360.com/whitecollar/articles/1353728/clear-exhibits-abrupt-goodbyes-life-as-a-zoom-juror.

6. Interview by Sarah Murray with Sean O'Donnell, Chief Criminal Judge, King County, Washington (Mar. 2021) [hereinafter Interview with Judge O'Donnell].

7. Salvatore, *supra* note 5.

8. *Id.*

influence goes down when they are separated, and people will tend to reach their own individual conclusions.[9]

Conducting Online Searches While Deliberating. The legitimate concern many have that jurors will try to access the internet for outside information during trial or while deliberating deserves attention. Even when admonished not to do so, we have seen mock jurors try to access the internet to "translate" a legal term in the jury instructions. Jurors' motives may be pure, but the danger associated with this outside influence increases when access to that information is so easy.

Influencing the Choice of Foreperson. At least one state court has recommended in its protocols that juries consider a foreperson with some technical proficiency, in part because the foreperson is responsible for communicating with the court if questions arise and when the jury has completed its deliberations: "While jurors have the absolute right to select their foreperson, it is not a bad idea to recommend that they either select a foreperson with technological prowess or designate another juror skilled in the use of modern technology who can assist with the digital process of remote deliberations."[10]

The Opportunities

Forging Civil Dialogue. In the Western District of Washington remote patent trial, juror David Lee found a speaking advantage using Zoom: "I've found that people are better about not talking over others on Zoom calls than they can be in person. That held true in this case."[11]

Discovering Increased Dialogue. Perhaps one surprise of the remote deliberation process is the trend revealed by our simulation work on behalf of the Arizona Superior Court. The majority of jurors reported that they felt the remote forum *increased* their willingness to speak during deliberations in comparison with in-person jury deliberations, as follows:[12]

This finding is meaningful. The individual voice of the juror in the deliberative process is sacrosanct. The fact that the remote trial environment amplifies that voice is telling. In July 2020, the Online Courtroom Project ran a remote jury trial demonstration over the course of two days. As our summary report concludes, our mock jurors had a similar experience:

---

9. Discussion with Ricky Raven, Litigation Partner, Reed Smith LLP (Mar. 22, 2021).

10. Riverside Superior Court, Protocol for Remote Jury Trials (2021), https://riverside.courts.ca.gov/generalinfo/attylitigants/remote-civil-jury-trials-protocol.pdf.

11. Salvatore, *supra* note 5.

12. *See* Gates et al., *supra* note 2.

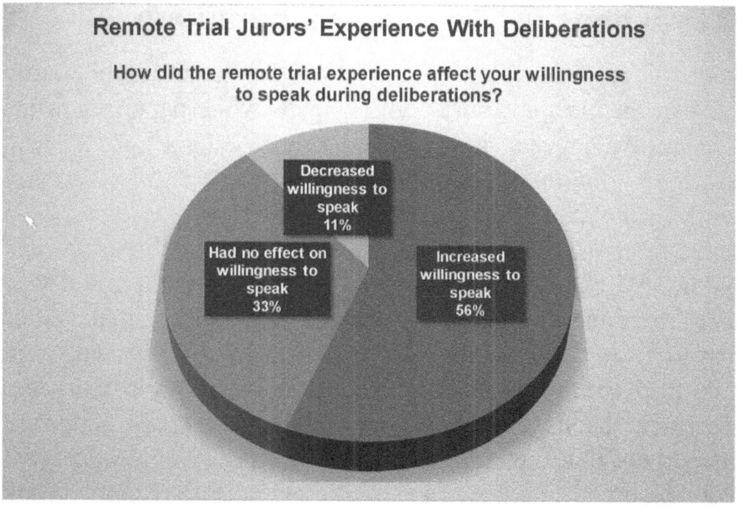

Figure 19.2

At the end of the trial, deliberations proceeded smoothly. Jurors had a vigorous discussion, with most of the jurors speaking up. Jurors felt comfortable talking with one another, and they indicated in debriefing discussions that they found it easy to use Zoom and the technology did not hinder their connection.[13]

Protecting the Spirit of Deliberative Debate. The Online Courtroom Project's demonstration trial also illuminated an important protection—that of fluid debate:

> The panel followed the same sequence that we generally see in in-person mock jury deliberations involving "around-the-table" sharing of individual opinions. First, mock jurors readily elected a foreperson, who immediately turned to the discussion of the negligence element of the case. They were able to easily speak, were easily recognized in the virtual room through either the Speaker or Gallery view by Zoom's color surround of the active speaker. Jurors were able to be heard and managed turn-taking during their deliberations. The resulting deliberation seemed to be as fluid as it would have been in person.[14]

13. Online Courtroom Project, Online Jury Trials: Summary and Recommendations (2020), https://www.onlinecourtroom.org/demonstration-report.
14. *Id.*

Best Practices for Remote Deliberations

Many of the recommended best practices for remote deliberations are reflected in the Online Courtroom Project's white paper regarding best practices for the entire trial.[15]

Lead jurors in an exemplar conversation to get to know one another. All newly formulated juries encounter an adjustment period as they begin to engage in conversation with one another. We know that online juries face additional challenges because the video format is initially foreign to some jurors and the fact of participating from home makes it easy to move away from the camera during breaks. An interview with a number of jurors from the first Oregon state court jury to participate in an online trial revealed the need for overt modeling. Despite the judge's strong encouragement that they talk to one another when in the breakout jury room during breaks, no one took that initiative, leaving them to talk with one another for the first time during deliberations.[16] To break the ice, consider giving jurors a topic to discuss (like going around the virtual room and saying a bit about their expectation of how this remote jury process will go) or model an ice-breaking conversation yourself. When you disclose something about your own background, jurors will appreciate learning a bit about you and will see that leadership as opportunity to open up more with their peers.

Create daily bonding opportunities before deliberations.
Give jurors daily opportunities to meet in a breakout room designated as the "jury room" and ensure that they move to the jury room together anytime counsel needs to be heard outside the presence of the jury. As a few courts have done, ask jurors to arrive 30 minutes ahead of court time each day to gather as a group before formally starting the day. Actively encourage jurors to use that time to talk about themselves or about other informal topics (while, of course, steering clear of case discussions). If any jurors tend to opt out of this time or give short shrift to the full time allotted, remind those jurors of the importance of spending time with their fellow jurors. When giving jurors breaks throughout the day, consider expanding the time to enable them to take a personal break first at home as well as time to gather with one another with frequency.

---

15. Richard Gabriel, Online Courtroom Project & National Institute for Trial Advocacy, The Online Courtroom and the Future of Jury Trials (2021).

16. Confidential Juror Interviews, *supra* note 1.

Recognize the importance of seating alternate jurors.
While juror attrition due to conventional reasons such as travel difficulties will abate in the online trial, new challenges demand seating alternate jurors. Some may face technical challenges that they cannot overcome, either because of their setup or because they lack the technical ability to overcome minor issues, even with the help of a technical bailiff. Technology aside, a juror may also fall ill or have an unexpected personal emergency, requiring that an alternate fill that empty seat.

Encourage jurors to enhance their "deliberation tech setup."
The deliberation phase marks a juror's shift from silent absorber to active participant. This shift warrants a technology setup that literally shines more light on each juror. We recommend two enhancements to ensure all jurors can see and be seen: (1) ample lighting on the juror's face either through lamplight in front of the juror's screen or ring lighting in the same location, and (2) wired earbuds to focus the juror's voice and to reduce ambient noise.

Plan ahead for technical disruptions.
Each juror needs a backup plan in the event he or she loses a connection to the group discussions. Instruct jurors that if one drops off, all case discussions must cease immediately until the dropped juror reconnects. Ensure that all jurors have telephonic access to the technical bailiff at all times during deliberations to help the dropped juror get back online as quickly as possible.

Influence jurors' views of their fellow jurors during deliberations.
Depending on the court and the video platform, jurors may have options for viewing their fellow jurors during deliberations. Most common to platform viewers, they could choose "active speaker view" where the talker dominates the screen or "gallery view" where all participants are present whether or not they are speaking. If jurors have the choice, urge them to keep the option set to "gallery view." This consistent view should promote greater identity as a group and the ability to watch one another for facial reactions to the conversation. In addition, "gallery view" should help jurors with verbal turn-taking during deliberations. One Oregon remote juror volunteered that a fellow juror could not access "gallery view" because he was participating from his tablet. As a result, he unintentionally talked over his fellow jurors because he could not see all of them like jurors working from computers could.[17]

Vary how you address outside influences.
Since jurors will have been repeatedly admonished during trial to keep their physical environment quiet and solitary, the "deliberation

17. *Id.*

admonishment" will be more effective if it is slightly different and if the *reasoning* behind the admonishment is clear. Long before the online trial came into existence, one judge noted, "If jurors are going to be asked to sacrifice some of their personal freedom and forgo their case-specific emailing, texting, blogging, instant messaging, and social networking for the duration of their service, they are entitled to a clear and thoughtful explanation of the reason."[18]

Focus on the *consequences* of even well-intentioned violations of two key admonishments. First, ask jurors to protect against an "uninvited juror" or someone in the periphery who is off-screen but still a potential source of influence on the juror who has allowed that person to be present during deliberations. Put another way, jurors should show trust in one another by limiting their case discussions only with one another, including during breaks or evenings when it might be tempting to rehash things with a nearby loved one. That external person cannot have the shared perspective that the other jurors have. Second, vary your reminder to avoid doing internet research about the case, the parties, or terminology even though it may be the juror's good intention to contribute clarification to the process. That internet "clarification" may be wrong and could unravel the hard work of the court and the parties who imparted knowledge during trial.

Give the foreperson a "hotline" to the court.

To ensure jurors can continue with deliberations as expeditiously as possible when an issue or question arises, ensure that the foreperson has both email and telephonic access to the court. In addition, ask the foreperson to notify the court when going offline for a break or for the day and then again when the jury goes back online.

Encourage the jury to ensure a voice for each juror.

Some jurors may feel intimidated by the video platform medium. Our simulations with remote jurors revealed some reticence on the part of a minority of jurors to being seen on-screen over a protracted period of time. For these few jurors, the on-screen experience can feel like a public speaking setting. Encourage all jurors to look out for one another and to invite comment from everyone. Empowering all jurors to take on this responsibility is preferable to simply making it the job of the foreperson.

---

18. Amy J. St. Eve & Michael A. Zuckerman, *Ensuring an Impartial Jury in the Age of Social Media*, 11 Duke L. & Tech. Rev. 1, 28 (2012).

Create interactive access to exhibits.
Jurors from recent remote jury trials[19] have expressed the importance of access to electronic exhibits for all jurors and have also voiced a strong preference for the ability of *each* juror to be able to interact with those exhibits using highlighting and markup tools like those provided in PDFs.

Try to avoid "Zoom fatigue" during protracted discussions.
The fatigue participants can feel while on video platforms is real. Encourage jurors to agree on the timing of frequently scheduled breaks rather than try to "tough out" a longer duration between breaks. Remind jurors that each individual may fatigue differently and that more frequent breaks should help everyone's concentration.

Ensure visible voting modes if the jury is so inclined.
In a typical jury room, a jury who opts for "visible voting" versus secret ballots can use a show of hands. In a Zoom room, a juror's "half-hearted" hand raise may be harder to see or may simply be an involuntary gesture. Encourage jurors to hold up something like a bright yellow sticky note when voting or to have the foreperson read off juror's numbers to confirm which hands were raised. Notably, some jurors in an online trial in King County, Washington, expressed a desire to have kept their individual votes more anonymous to alleviate peer pressure but ultimately could not figure out an online method for enacting a secret ballot mode.[20]

Have a verdict form completion and "chain of custody" plan.
Ideally, all jurors should have access to an electronic PDF of the verdict form that they can review during deliberations, while the foreperson would be the only one to have a form that can be completed online. If a "fillable" verdict form is not an option, consider the approach an Oregon remote jury foreperson described from their deliberation experience:

> I received a pdf of the verdict form and the method I used was I just held it up in front of the screen [to show my fellow jurors.] I printed it out and filled it out by hand. It was a pretty simple verdict form and as I recall, there was a place for me to print my name and a space to hand write in the dollar amount of the award and a place for me to sign. I showed the completed verdict form to the jurors and I scanned the completed verdict form and emailed it back to the clerk. The clerk then printed out the signed verdict, folded it up, put it in an envelope and physically handed it to the judge. The judge brought us back into the [online] courtroom

19. Salvatore, *supra* note 5.
20. Interview with Judge O'Donnell, *supra* note 6.

> and the judge asked if we reached a verdict and he took the envelope and he opened it in front of us and read what we awarded to the plaintiff and asked all of the jurors collectively if they agreed that was the amount we agreed to in deliberation.[21]

And, finally, allow jurors finality.

At least one remote juror expressed regret that the completion of trial meant an abrupt sign-off from the platform.[22] Consider allowing jurors time to chat online after they have been officially dismissed from jury duty. Their trial experience may have been remote but their connection with one another is typically as real in the two-dimensional forum as it has been in the three-dimensional courtroom.

21. Confidential Juror Interviews, *supra* note 1.
22. Salvatore, *supra* note 5.

Chapter 20

The Future of Online Litigation

Richard Gabriel

When I was a child, my mother had a private family law practice out of our house, so I spent a lot of time at local libraries. And even though I have a special reverence for those places where I spent countless hours as a kid, it was the books and not the building that made them special. When my mother got appointed to the bench, one of her courtrooms was a small prefab bungalow in the parking lot of the Santa Monica courthouse, no different than a windowless middle school classroom. And every day that she showed up to work and every trial she conducted in her courtroom carried the same weight and solemnity as every trial I have ever done in a marble and mahogany federal building.

The pandemic changed much of what we had done in everyday life: the way we worked, traveled, socially interacted, recreated, and the way our kids went to school. And even though there had been a great deal of emphasis about "getting back to normal," this crisis allowed us to take a long, hard look at the way that lawyers and law firms practice, and the way the courts operate.

In 2021, Cushman and Wakefield's Global Legal Sector Advisory Group published their annual survey of hundreds of attorneys. In that poll, 84 percent of the respondents anticipate they will be spending a great deal more on remote or networking capabilities in the future. Additionally, 78 percent of firms expect associates to work remotely at least two days a week when the pandemic is over, with partners only anticipated to come into the office one to three days a week. Quinn Emanuel Urquhart & Sullivan, one of the largest litigation firms in the U.S., recently stated that it would allow any of its attorneys to "work from wherever they want."[1]

1. *Quinn Emmanuel Tells Its Lawyers: "Work from Anywhere,"* QUINN EMMANUEL TRIAL LAWYERS, https://www.quinnemanuel.com/the-firm/news-events/quinn-emanuel-tells-its-lawyers-work-from-anywhere/ (last visited Feb. 2, 2022).

During the pandemic, 62 percent of Americans worked from home compared to 25 percent the previous year.[2] According to these McKinsey & Company surveys, 80 percent said they enjoyed working at home, 41 percent said they felt more productive, and 28 percent said they felt as productive as when they were at work.[3] As a result, they found that 86 percent of the American workforce would like to work remotely at least one day a week[4] and a survey of federal workers and contractors found that nearly half would like to work remotely full time.[5]

Despite resistance, progress marches inexorably forward and technology is part of that change. Americans spend, on average, six and a half hours a day on a computer device[6] and screens are part of our everyday landscape in stores and restaurants. Law firms are in heated competition with or are using alternative legal service providers (ALSPs) who employ technology such as artificial intelligence (AI) for e-discovery, document review, and legal analysis. Instead of creating a crisis in the legal community, the pandemic may have just accelerated changes that were already occurring at a slower pace. Hybrid work models are the future and technology upgrades to all workplace systems will become necessary in all work sectors.

The coronavirus pandemic challenged us not only to be flexible and resourceful in our "new normal" work practices, but also provided us with the opportunity to look at how we can improve law practices and court operations in the future. Here are a few areas of online or remote litigation and trial practice that are likely to continue after the pandemic as well as additional reforms that may improve our justice system. As former U.S. District Court Judge Jeremy Fogel (ret.), who is director of the Berkeley Judicial Institute, has said, "History teaches us that a crisis

2. Megan Brenan, *U.S. Workers Discovering Affinity for Remote Work*, Gallup, Apr. 3, 2020, https://news.gallup.com/poll/306695/workers-discovering-affinity-remote-work.aspx.

3. Brodie Boland et al., *Reimagining the Office and Work Life after COVID-19*, McKinsey & Co., June 8, 2020, https://www.mckinsey.com/business-functions/organization/our-insights/reimagining-the-office-and-work-life-after-covid-19.

4. Andrea Alexander et al., *What Employees Are Saying about the Future of Remote Work*, McKinsey & Co., Apr. 1, 2021, https://www.mckinsey.com/business-functions/organization/our-insights/what-employees-are-saying-about-the-future-of-remote-work.

5. Catharina Wrede Braden et al., *Reimagining US Federal Work for the Postpandemic World*, McKinsey & Co., Apr. 6, 2021, https://www.mckinsey.com/industries/public-and-social-sector/our-insights/reimagining-us-federal-work-for-the-postpandemic-world.

6. *Screen Time: UK vs US vs The Rest of the World Compared*, Bus. Fibre, July 15, 2019, https://businessfibre.co.uk/screen-time/.

often can be the catalyst of innovations that endure long after the crisis itself has ended."[7] As an example, Michigan has pioneered an online dispute resolution platform called MI-Resolve that actually allows state residents, at no cost, to resolve disputes with or without a mediator on a phone, tablet, or laptop without going to court.[8]

Based on our experience from the pandemic, and while every jurisdiction, courthouse, courtroom, law practice, and case will be different, this is my opinion of which general remote litigation practices will remain after the pandemic. While the courts have been reluctant to tackle the thorny issue of whether remote technology meets Sixth Amendment constitutional standards, the following areas will be employed in a civil litigation context, in either a "hybrid" form with some participants live in court while others appear remotely by videoconferencing technology, or some proceedings will be completely remote:

- Client meetings
- Administrative hearings, motions, or court conferences
- Depositions
- Mediations
- Arbitrations
- Bench trials
- Jury trials
- Media access to court proceedings
- Legal education and conferences

Efficiency and Economy

One of the main reasons given for the "vanishing jury trial" is the ever-increasing cost of litigation. One of the issues that needs to be looked at more extensively in the future is how remote technology and procedures could make litigation less costly, and thus more accessible for a wider population. Federal Rule of Civil Procedure 1 mandates that the court strike a balance between justice, efficiency, and cost.[9] Online litigation practices provide a number of cost-effective measures. First, the courts encountered

7. *Expanding Access to the Federal Courts: The Unexpected Opportunity Presented by the COVID-19 Pandemic* (June 25, 2020) (statement of Hon. Jeremy Fogel (ret.), Exec. Dir., Berkeley Jud. Inst.), https://docs.house.gov/meetings/JU/JU03/20200625/110837/HHRG-116-JU03-Wstate-FogelJ-20200625.pdf.

8. Michigan Courts, *Resolve a Dispute Online with MI-Resolve*, https://www.courts.michigan.gov/administration/offices/office-of-dispute-resolution/mi-resolve/ (last visited Nov. 6, 2021).

9. Fed. R. Civ. P. 1.

an immense backlog of cases as a result of months, if not more than a year, of court closures. As a result, judges and lawyers alike experienced the efficiency of online hearings and depositions. In online court proceedings, there are no traffic delays, long commutes, difficulty parking, waiting for elevators, or bad cafeteria coffee. There is quicker and easier access to forgotten or unanticipated documents. Judges who have questions on briefing or motions can simply send an email to the attorneys and set up a videoconferencing call to address the issue rather than set a new hearing date.

In multidistrict litigation (MDL) or matters where attorneys, parties, witnesses, clients, and claims adjusters would normally have to travel across the country to attend depositions, hearings, and trial, these parties would now simply open up a laptop and click on a videoconference link. Flight delays, traffic accidents, and bad weather creating unexpected changes in the scheduling of witnesses, or the cause of missing jurors, would no longer become an issue. Without travel and its attendant uncertainty, trial scheduling is easier and less expensive. In an online trial forum, direct and cross-examination of a witness may also be able to be prerecorded to accommodate witness schedules or a particular sequence of evidence that the attorneys wish to present. Depending on scheduling and timing, even opening statements and closing arguments can be prerecorded. More importantly, remote proceedings can and should be more streamlined. Attorneys and clients would need to take a good hard look at their cases to see what evidence really *needs* to be presented; extraneous opinion and redundant testimony edited to provide fact finders with only the necessary elements of the case. We have a habitual impatience with technology—we want pages and documents to download quicker and want people to get to the point in our everyday online meetings. This also translates in the legal setting to hearings, depositions, and trial. Because of the significant backlog in cases caused by the pandemic, the prerecording of segments of a trial or even entire trials can help with less complicated matters that can fit into a summary or expedited trial model offered by most states.

In the legal profession, we all live busy and stressful lives. Aside from the time and cost benefits of remote litigation practices, if videoconferencing technology can make our lives a little less stressful and allow us a little more time with our loved ones, that by itself can be worth the initial discomfort of learning new skills and practices.

Technology

While courts have returned to in-person proceedings, courtroom technology will and must evolve to meet changing litigation demands. As mentioned previously in Chapter 13, the Coronavirus Aid, Relief, and Economic Security (CARES) Act[10] has paid for hundreds of courtroom technology upgrades and funding may still be available in the future for additional courtroom upgrades. An integrated courtroom technology solution provides a number of advantages for courts and counsel, whether they are conducting proceedings in person or online. First, assisted listening devices are more easily integrated into these systems for participants who may be hard of hearing. Additionally, while courts may not have the budget for traditional court interpreters for foreign language jurors, a remote system may allow an online interpreter to listen to or watch a proceeding and give a simultaneous translation for a juror or jurors to more easily comprehend the evidence in a case. An integrated technology system also allows for greater evidence control by the judge and clerk if the court sets up a procedure where an exhibit is sent electronically to the bench, is ruled on, and then shown to the remote or in-court jury if allowed. Adam Lofredo, who is director of the Justice Division for Conference Technologies, a company that has been providing technology upgrades for hundreds of courtrooms across the United States, believes that the key to all courtroom technology solutions is clarity and ease of use: "Technology has to create a positive experience for participants in a courtroom setting. So, the technology choices that the courts make should be geared toward creating that positive experience. At the same time, the technology needs to integrate seamlessly into the courts' daily process."[11]

In practical terms, will the participants be able to easily see or hear the evidence? Will the attorneys, the judge, and the courtroom staff be able to use the technology relatively easily to present their cases, make their motions and rulings, and decide the matters they are called on to judge? As with any new practice, there will be a learning curve. But note how many in the legal profession have developed proficiency in a matter of months using videoconferencing technology, with its "share screens" and "chat boxes."

10. The CARES Act directs the federal rule makers to "consider rule amendments . . . that address emergency measures." Pub. L. No. 116-136, 134 Stat. 281 (2020).

11. Interview with Adam Lofredo, Director of Justice Division, Conference Technologies, Inc. (Apr. 8, 2020).

Access

Courtroom technology and online proceedings also provide greater access to the courts for a variety of constituents. As described in Chapter 8 of this book, lower socioeconomic jurors and minority populations currently provide the largest no-show rates for jury service. This can be caused by economic hardship, transportation issues, access to childcare, fear of missing work, or the perception that they cannot participate because of poor English comprehension. Some minority populations also have a negative association with the court system and simply do not want to serve because of their prior poor experiences. According to the Sentencing Project, Black men and women are incarcerated in state prisons at a rate more than five times the imprisonment of Whites. In some states, that ratio is more like 10:1. In 12 states, more than half the prison population is Black and in Maryland, the Black prison population is 72 percent. In Oklahoma, one in 15 Black males over 18 years old in the state is in prison. Latino imprisonment is also higher, 1.4 times that of the White population and, in some states, it is three or four times the rate of White prisoners.[12] Some state courts may struggle with minority juror representation simply because minority populations may associate the courts with incarceration, deportation, or the perception of an unequal justice system.

Remote technology can also provide greater access to pro se defendants, parties in child custody hearings who cannot take off work for extended periods of time, or litigants who have to take care of children, elderly, or disabled family members. This type of access also helps disabled and hard of hearing populations, or participants that may have special medical needs that necessitate they be close to home, whether they are parties, jurors, or witnesses. Additionally, jurors who live in outlying federal court counties or any attorney or witness who lives a significant distance from a courthouse in a rural area can have easier access to the courts. This could also have an effect on Federal Rule of Civil Procedure 45,[13] which requires subpoenas of nonparties to compel compliance within 100 miles of the nonparty, to avoid overburdening the nonparty with distant travel. If the proceedings are remote, there is no burden.

One of the main complaints during the pandemic is that incarcerated defendants could not meet with their lawyers to prepare for trial.

12. Ashley Nellis, The Sentencing Project, The Color of Justice: Racial and Ethnic Disparity in State Prisons (2021), https://www.sentencingproject.org/publications/color-of-justice-racial-and-ethnic-disparity-in-state-prisons/.

13. Fed. R. Civ. P. 45.

Even after the pandemic, remote technology could create greater access to counsel for in-custody defendants awaiting trial or appeal.

According to the testimony of Michigan Supreme Court Chief Justice Bridget McCormack:

> Since April 1, [2020], judges and other court officers have held well over 50,000 Zoom meetings and are approaching 350,000 hours of hearings held online. And to maintain public access to court proceedings, virtual hearings conducted by Zoom are being livestreamed to YouTube. To make public access to those livestreams easy, our tech team set up a Virtual Courtroom Directory with a clickable map so that users can click on their county, find their judge and watch. This directory has been used more than 25,000 times in the past month alone.[14]

In July of 2021, the California Judicial Council approved funding to continue providing remote access to court proceedings, identifying it as an "equal access to justice issue." Justice Marsha Slough, who was chair of the Chief Justice's Ad-Hoc Workgroup on Post-Pandemic Initiatives, stated, "We simply would be wrongheaded after all we've heard so far to unplug remote proceedings as a tool to be used in accessing justice in our courts."[15] In responding to the ad-hoc workgroup's recommendations, Chief Justice Cantil-Sakauye added that remote access has been a long-term goal for the judicial branch and stated, "What it has come down to in your moving remarks to me is that remote access to justice, hands down, serves our most vulnerable."[16]

Remote access to our justice system increases the participation of all stakeholders in the system.

The Great Equalizer

While the ceremony of trials carries a solemnity and an authority that conveys importance, it also creates barriers and may actually hinder decision-making. Witnesses and jurors often feel intimidated, sometimes inadequate, and some are downright scared in a courtroom environment. This can create an added tension that interferes with the cognitive clarity

14. *Federal Courts during the Covid-19 Pandemic Hearing*, *supra* note 7 (statement of Hon. Bridget M. McCormack, Chief Justice, Michigan Supreme Court).

15. News Release, California Courts, Council Approves Funding Allocation for Trial Courts (July 9, 2021).

16. *Id.*

of a witness or juror. But when they are in their own homes or in familiar surroundings, they tend to be more comfortable and candid. With remote presentations, there may be less formality, but there is a greater sense of equality in stature between the judge, jury, attorneys, and witnesses. Everyone looks pretty much the same: a small square on a screen. There is the inevitable wandering pet, interrupting child, or computer glitch. While these pauses may be momentary distractions from a proceeding, it also makes everyone *human*. Some judges who have used videoconferencing for sentencing hearings have said that they have had a different perspective on a defendant when they could see them on camera in their home environment interacting with a spouse, significant other, or children, rather than the sterile formality of the courtroom.

Pro se defendants also may feel less intimidated or outmatched by opposing counsel. Women and minorities who traditionally have not had as much of a voice as their White male counterparts may feel freer to share their opinions. A larger juror cannot impose their physical size to gain more influence in deliberations. Online proceedings also tend to promote greater civility in discourse. As some judges have said this year, it is a "cool medium" where bluster and aggressive hostility does not play well. In fact, videoconferencing can make it easier to focus on substantive evidence rather than the personalities in the case.

There is also a kind of intimacy with the camera in a remote interaction that is not present in court. In court, an attorney is speaking to a group, which can lend itself to a lecture-type of presentation and, if unchecked, bombast. In a remote presentation, you are speaking to one person, because every person is alone listening in their home and you are speaking to them and them alone. The ability to connect on a human level, contrary to popular belief, does not have to be minimized. Seeing people in their homes and even their work environments gives a more genuine view of them and prompts *us* to be more authentic in our interactions. Does that make us less professional? No. Because our professionalism lies in our ability to analyze, to argue, to advocate, and to communicate.

Conclusion

The National Center for State Courts has been conducting polling on public attitudes about the courts since 2012. In their 2021 survey,[17] there

17. Memo from GBAO, to National Center for State Courts, re: 2021 State of the State Courts—National Survey Analysis (Oct. 29, 2021).

was a marked increase in the use of remote proceedings for traffic tickets, consumer debt, small claims, landlord/tenant disputes, divorce, and child custody hearings. For most of these, respondents in the survey stated a preference (sometimes a strong preference) for the remote proceeding, with divorce and child custody understandably preferring in-person proceedings. The survey also indicates that respondents would be likely to appear remotely for a mediation, arbitration, or trial, and to report and serve on a jury. A majority of the respondents agreed that the courts should continue remote proceedings because it would allow them to hear more cases, resolve cases more quickly, and make it easier for people to participate in court proceedings.

The Pew Charitable Trusts also noted the remarkable adoption of technology in court proceedings in 2020 through 2021,[18] highlighting that Texas courts held 1.1 million remote proceedings in that time period, even though it had never previously done so. In a two-month period in 2020, Michigan held 35,000 video hearings that it had previously never conducted. The report also noted that online proceedings did prove more difficult for those without lawyers, those without access to technology, those with English language deficiency, and disabled populations, and these issues should be addressed in future discussions about implementing remote proceedings. However, the report demonstrated that despite these difficulties, there was a marked increase in participation in court proceedings as a result of applied technology solutions. In Arizona,[19] Texas,[20] New Jersey,[21] and Michigan,[22] courts also saw a decline in default judgments, and a decrease in failure-to-show rates for criminal, debt collection, eviction, family law, and child welfare cases. In the future, increased government infrastructure spending may also alleviate some of the internet access issues.

---

18. Pew Charitable Trusts, How Courts Embraced Technology, Met the Pandemic Challenge, and Revolutionized Their Operations (2021), https://www.pewtrusts.org/-/media/assets/2021/12/how-courts-embraced-technology.pdf.

19. COVID-19 Continuity of Court Operations during a Public Health Emergency Workgroup, Arizona Supreme Court, Post-Pandemic Recommendations (2021), https://www.americanbar.org/content/dam/aba/administrative/judicial/2021-az-post-pandemic-rec.pdf.

20. E.G. Thornburg, *Observing Online Courts: Lessons from the Pandemic*, 54 Fam. L.Q. 208–09 (2021), https://www.americanbar.org/groups/family_law/publications/family-law-quarterly/volume-54/issue-3/observing-online-courts-lessons-the-pandemic/.

21. *Will Remote Hearing Improve Appearance Rates?*, Nat'l Center for St. Cts., May 13, 2020, https://www.ncsc.org/newsroom/at-the-center/2020/may-13.

22. *Id.*

The Pew report also highlighted efforts of national judicial groups such as the Conference of Chief Justices (CCJ) and the Conference of State Court Administrators (COSCA), which demonstrated a commitment to "a more open, equitable, and efficient civil legal system."[23] These two organizations proposed six principles to guide the use of technology in post-pandemic courts.[24] These include ensuring principles of due process, procedural fairness, transparency, and equal access, along with focusing on the user experience, prioritizing court technology, embracing flexibility and willingness to adapt, adopting remote technology where practicable, and taking an open and data-driven approach to move court processes forward. In June of 2021, a survey of magistrate, trial judges, and appellate justices indicated that a majority expected online proceedings to become a future norm for the courts.[25]

While this book has extensively discussed the benefits of remote litigation, it does not advocate this medium as a substitute for in-person proceedings. Rather, it is a resource to be used, as needed, depending on the case and the proceeding. Given that these remote litigation practices are relatively new, we should not be afraid to study, to experiment, and to research how these types of proceedings affect the different constituents in our court system, and to develop best practices based on research and training. In the appendix and resource block of this book, you will find exemplar surveys, interview questionnaires, and other resources to initiate this process.

From infectious agent identification to license, it took 16 years to develop a hepatitis B vaccine. It took nine and a half months to develop the COVID-19 vaccine and the technology used to develop these vaccines so quickly will no doubt assist in future medical research. A crisis can be a call to action to remedy and restore our way of life. But it can also be a call to explore, to innovate, and to improve existing systems. Practitioners, court systems, and academics should take the opportunity

---

23. *Id.*

24. National Center for State Courts, Guiding Principles for Post-Pandemic Court Technology: A Pandemic Resource from CCJ/COSCA (2020), https://www.srln.org/system/files/attachments/Guiding%20Principles%20for%20Post%20Pandemic%20Court%20Technology%20%28CCJ%3ACOSCA%202020%29.pdf.

25. Gina Jurva, Thomson Reuters, The Impacts of the COVID-19 Pandemic on State & Local Courts (2021). *See also* Allie Reed & Madison Alder, *Zoom Courts Will Stick Around as Virus Forces Seismic Change*, Bloomberg L., July 30, 2020, https://news.bloomberglaw.com/us-law-week/zoom-courts-will-stick-around-as-virus-forces-seismic-change.

to conduct research, start conversations, develop networks to exchange ideas, and examine a number of issues and practices raised in this book. Namely:

- What features of the online legal process work as well, better, or worse than their in-person counterparts?
- How do online legal practices affect the overall cost of litigation?
- How do online legal practices affect the efficiency of the litigation process, including discovery, settlement, and trial?
- Can online litigation practices be employed in criminal cases without violating the constitutional rights of the defendant?
- What specific practices in the online legal process serve to include or exclude different stakeholders and affect access to and representation in the courts?
- How do online presentations affect the clarity of court presentations and fact finder attention, retention, and use of presented evidence and arguments?
- How do online presentations of evidence, testimony, and arguments affect the evaluation of the credibility of the evidence, witnesses, and attorneys, as well as fact finder decision-making?

These are only a few of the important questions raised by remote litigation practices. The authors of this book and the Online Courtroom Project look forward to how the courts, law firms, and other constituents of the legal system engage in discussing, studying, and utilizing these evolving forms in order to facilitate a more equitable and fair administration of justice.

Appendix A

AMERICAN BAR ASSOCIATION

STANDING COMMITTEE ON ETHICS AND PROFESSIONAL RESPONSIBILITY

Formal Opinion 498 **March 10, 2021**

Virtual Practice

The ABA Model Rules of Professional Conduct permit virtual practice, which is technologically enabled law practice beyond the traditional brick-and-mortar law firm.[1] *When practicing virtually, lawyers must particularly consider ethical duties regarding competence, diligence, and communication, especially when using technology. In compliance with the duty of confidentiality, lawyers must make reasonable efforts to prevent inadvertent or unauthorized disclosures of information relating to the representation and take reasonable precautions when transmitting such information. Additionally, the duty of supervision requires that lawyers make reasonable efforts to ensure compliance by subordinate lawyers and nonlawyer assistants with the Rules of Professional Conduct, specifically regarding virtual practice policies.*

I. Introduction

As lawyers increasingly use technology to practice virtually, they must remain cognizant of their ethical responsibilities. While the ABA Model Rules of Professional Conduct permit virtual practice, the Rules provide some minimum requirements and some of the Comments suggest best practices for virtual practice, particularly in the areas of competence, confidentiality, and supervision. These requirements and best practices are discussed in this opinion, although this opinion does not address every ethical issue arising in the virtual practice context.[2]

II. Virtual Practice: Commonly Implicated Model Rules

This opinion defines and addresses virtual practice broadly, as technologically enabled law practice beyond the traditional brick-and-mortar law firm.[3] A lawyer's virtual practice often occurs when a lawyer at home or on-the-go is working from a location outside the office, but a lawyer's practice may be entirely virtual because there is no requirement in the Model Rules that a lawyer

---

[1] This opinion is based on the ABA Model Rules of Professional Conduct as amended by the ABA House of Delegates through August 2020. The laws, court rules, regulations, rules of professional conduct, and opinions promulgated in individual jurisdictions are controlling.

[2] Interstate virtual practice, for instance, also implicates Model Rule of Professional Conduct 5.5: Unauthorized Practice of Law; Multijurisdictional Practice of Law, which is not addressed by this opinion. *See* ABA Comm. on Ethics & Prof'l Responsibility, Formal Op. 495 (2020), stating that "[l]awyers may remotely practice the law of the jurisdictions in which they are licensed while physically present in a jurisdiction in which they are not admitted if the local jurisdiction has not determined that the conduct is the unlicensed or unauthorized practice of law and if they do not hold themselves out as being licensed to practice in the local jurisdiction, do not advertise or otherwise hold out as having an office in the local jurisdiction, and do not provide or offer to provide legal services in the local jurisdiction."

[3] *See generally* MODEL RULES OF PROFESSIONAL CONDUCT R. 1.0(c), defining a "firm" or "law firm" to be "a lawyer or lawyers in a partnership, professional corporation, sole proprietorship or other association authorized to practice law; or lawyers employed in a legal services organization on the legal department of a corporation or other organization." Further guidance on what constitutes a firm is provided in Comments [2], [3], and [4] to Rule 1.0.

have a brick-and-mortar office. Virtual practice began years ago but has accelerated recently, both because of enhanced technology (and enhanced technology usage by both clients and lawyers) and increased need. Although the ethics rules apply to both traditional and virtual law practice,[4] virtual practice commonly implicates the key ethics rules discussed below.

A. Commonly Implicated Model Rules of Professional Conduct

1. Competence, Diligence, and Communication

Model Rules 1.1, 1.3, and 1.4 address lawyers' core ethical duties of competence, diligence, and communication with their clients. Comment [8] to Model Rule 1.1 explains, "To maintain the requisite knowledge and skill [to be competent], a lawyer should keep abreast of changes in the law and its practice, *including the benefits and risks associated with relevant technology*, engage in continuing study and education and comply with all continuing legal education requirements to which the lawyer is subject." (*Emphasis added*). Comment [1] to Rule 1.3 makes clear that lawyers must also "pursue a matter on behalf of a client despite opposition, obstruction or personal inconvenience to the lawyer, and take whatever lawful and ethical measures are required to vindicate a client's cause or endeavor." Whether interacting face-to-face or through technology, lawyers must "reasonably consult with the client about the means by which the client's objectives are to be accomplished; . . . keep the client reasonably informed about the status of the matter; [and] promptly comply with reasonable requests for information. . . ."[5] Thus, lawyers should have plans in place to ensure responsibilities regarding competence, diligence, and communication are being fulfilled when practicing virtually.[6]

2. Confidentiality

Under Rule 1.6 lawyers also have a duty of confidentiality to all clients and therefore "shall not reveal information relating to the representation of a client" (absent a specific exception, informed consent, or implied authorization). A necessary corollary of this duty is that lawyers must at least "make reasonable efforts to prevent the inadvertent or unauthorized disclosure of, or unauthorized access to, information relating to the representation of a client."[7] The following non-

[4] For example, if a jurisdiction prohibits substantive communications with certain witnesses during court-related proceedings, a lawyer may not engage in such communications either face-to-face or virtually (e.g., during a trial or deposition conducted via videoconferencing). *See, e.g.*, MODEL RULES OF PROF'L CONDUCT R. 3.4(c) (prohibiting lawyers from violating court rules and making no exception to the rule for virtual proceedings). Likewise, lying or stealing is no more appropriate online than it is face-to-face. *See, e.g.*, MODEL RULES OF PROF'L CONDUCT R. 1.15; MODEL RULES OF PROF'L CONDUCT R. 8.4(b)-(c).

[5] MODEL RULES OF PROF'L CONDUCT R. 1.4(a)(2) – (4).

[6] Lawyers unexpectedly thrust into practicing virtually must have a business continuation plan to keep clients apprised of their matters and to keep moving those matters forward competently and diligently. ABA Comm. on Ethics & Prof'l Responsibility, Formal Op. 482 (2018) (discussing ethical obligations related to disasters). Though virtual practice is common, if for any reason a lawyer cannot fulfill the lawyer's duties of competence, diligence, and other ethical duties to a client, the lawyer must withdraw from the matter. MODEL RULES OF PROF'L CONDUCT R. 1.16. During and following the termination or withdrawal process, the "lawyer shall take steps to the extent reasonably practicable to protect a client's interests, such as giving reasonable notice to the client, allowing time for employment of other counsel, surrendering papers and property to which the client is entitled and refunding any advance payment of fee or expense that has not been earned or incurred." MODEL RULES OF PROF'L CONDUCT R. 1.16(d).

[7] MODEL RULES OF PROF'L CONDUCT R. 1.6(c).

exhaustive list of factors may guide the lawyer's determination of reasonable efforts to safeguard confidential information: "the sensitivity of the information, the likelihood of disclosure if additional safeguards are not employed, the cost of employing additional safeguards, the difficulty of implementing the safeguards, and the extent to which the safeguards adversely affect the lawyer's ability to represent clients (e.g., by making a device or important piece of software excessively difficult to use)."[8] As ABA Formal Op. 477R notes, lawyers must employ a "fact-based analysis" to these "nonexclusive factors to guide lawyers in making a 'reasonable efforts' determination."

Similarly, lawyers must take reasonable precautions when transmitting communications that contain information related to a client's representation.[9] At all times, but especially when practicing virtually, lawyers must fully consider and implement reasonable measures to safeguard confidential information and take reasonable precautions when transmitting such information. This responsibility "does not require that the lawyer use special security measures if the method of communication affords a reasonable expectation of privacy."[10] However, depending on the circumstances, lawyers may need to take special precautions.[11] Factors to consider to assist the lawyer in determining the reasonableness of the "expectation of confidentiality include the sensitivity of the information and the extent to which the privacy of the communication is protected by law or by a confidentiality agreement."[12] As ABA Formal Op. 477R summarizes, "[a] lawyer generally may transmit information relating to the representation of a client over the Internet without violating the Model Rules of Professional Conduct where the lawyer has undertaken reasonable efforts to prevent inadvertent or unauthorized access."

3. Supervision

Lawyers with managerial authority have ethical obligations to establish policies and procedures to ensure compliance with the ethics rules, and supervisory lawyers have a duty to make reasonable efforts to ensure that subordinate lawyers and nonlawyer assistants comply with the applicable Rules of Professional Conduct.[13] Practicing virtually does not change or diminish this obligation. "A lawyer must give such assistants appropriate instruction and supervision concerning the ethical aspects of their employment, particularly regarding the obligation not to disclose information relating to representation of the client, and should be responsible for their work product."[14] Moreover, a lawyer must "act competently to safeguard information relating to the representation of a client against unauthorized access by third parties and against inadvertent

---

[8] MODEL RULES OF PROF'L CONDUCT R. 1.6 cmt. [18].
[9] MODEL RULES OF PROF'L CONDUCT R. 1.6 cmt. [19].
[10] *Id.*
[11] The opinion cautions, however, that "a lawyer may be required to take special security precautions to protect against the inadvertent or unauthorized disclosure of client information when required by an agreement with the client or by law, or when the nature of the information requires a higher degree of security." ABA Comm. on Ethics & Prof'l Responsibility, Formal Op. 477R (2017).
[12] MODEL RULES OF PROF'L CONDUCT R. 1.6 cmt. [19].
[13] MODEL RULES OF PROF'L CONDUCT R. 5.1 & 5.3. *See, e.g.*, ABA Comm. on Ethics & Prof'l Responsibility, Formal Op. 467 (2014) (discussing managerial and supervisory obligations in the context of prosecutorial offices). *See also* ABA Comm. on Ethics & Prof'l Responsibility, Formal Op. 483 n.6 (2018) (describing the organizational structures of firms as pertaining to supervision).
[14] MODEL RULES OF PROF'L CONDUCT R. 5.3 cmt. [2].

or unauthorized disclosure by the lawyer *or other persons who are participating in the representation of the client or who are subject to the lawyer's supervision*."[15] The duty to supervise nonlawyers extends to those both within and outside of the law firm.[16]

B. Particular Virtual Practice Technologies and Considerations

Guided by the rules highlighted above, lawyers practicing virtually need to assess whether their technology, other assistance, and work environment are consistent with their ethical obligations. In light of current technological options, certain available protections and considerations apply to a wide array of devices and services. As ABA Formal Op. 477R noted, a "lawyer has a variety of options to safeguard communications including, for example, using secure internet access methods to communicate, access and store client information (such as through secure Wi-Fi, the use of a Virtual Private Network, or another secure internet portal), using unique complex passwords, changed periodically, implementing firewalls and anti-Malware/Anti-Spyware/Antivirus software on all devices upon which client confidential information is transmitted or stored, and applying all necessary security patches and updates to operational and communications software." Furthermore, "[o]ther available tools include encryption of data that is physically stored on a device and multi-factor authentication to access firm systems." To apply and expand on these protections and considerations, we address some common virtual practice issues below.

1. Hard/Software Systems

Lawyers should ensure that they have carefully reviewed the terms of service applicable to their hardware devices and software systems to assess whether confidentiality is protected.[17] To protect confidential information from unauthorized access, lawyers should be diligent in installing any security-related updates and using strong passwords, antivirus software, and encryption. When connecting over Wi-Fi, lawyers should ensure that the routers are secure and should consider using virtual private networks (VPNs). Finally, as technology inevitably evolves, lawyers should periodically assess whether their existing systems are adequate to protect confidential information.

---

[15] MODEL RULES OF PROF'L CONDUCT R. 1.6 cmt. [18] (emphasis added).

[16] As noted in Comment [3] to Model Rule 5.3:

> When using such services outside the firm, a lawyer must make reasonable efforts to ensure that the services are provided in a manner that is compatible with the lawyer's professional obligations. The extent of this obligation will depend upon the circumstances, including the education, experience and reputation of the nonlawyer; the nature of the services involved; the terms of any arrangements concerning the protection of client information; and the legal and ethical environments of the jurisdictions in which the services will be performed, particularly with regard to confidentiality. See also Rules 1.1 (competence), 1.2 (allocation of authority), 1.4 (communication with client), 1.6 (confidentiality), 5.4(a) (professional independence of the lawyer), and 5.5(a) (unauthorized practice of law).

[17] For example, terms and conditions of service may include provisions for data-soaking software systems that collect, track, and use information. Such systems might purport to own the information, reserve the right to sell or transfer the information to third parties, or otherwise use the information contrary to lawyers' duty of confidentiality.

2. Accessing Client Files and Data

Lawyers practicing virtually (even on short notice) must have reliable access to client contact information and client records. If the access to such "files is provided through a cloud service, the lawyer should (i) choose a reputable company, and (ii) take reasonable steps to ensure that the confidentiality of client information is preserved, and that the information is readily accessible to the lawyer."[18] Lawyers must ensure that data is regularly backed up and that secure access to the backup data is readily available in the event of a data loss. In anticipation of data being lost or hacked, lawyers should have a data breach policy and a plan to communicate losses or breaches to the impacted clients.[19]

3. Virtual meeting platforms and videoconferencing

Lawyers should review the terms of service (and any updates to those terms) to ensure that using the virtual meeting or videoconferencing platform is consistent with the lawyer's ethical obligations. Access to accounts and meetings should be only through strong passwords, and the lawyer should explore whether the platform offers higher tiers of security for businesses/enterprises (over the free or consumer platform variants). Likewise, any recordings or transcripts should be secured. If the platform will be recording conversations with the client, it is inadvisable to do so without client consent, but lawyers should consult the professional conduct rules, ethics opinions, and laws of the applicable jurisdiction.[20] Lastly, any client-related meetings or information should not be overheard or seen by others in the household, office, or other remote location, or by other third parties who are not assisting with the representation,[21] to avoid jeopardizing the attorney-client privilege and violating the ethical duty of confidentiality.

4. Virtual Document and Data Exchange Platforms

In addition to the protocols noted above (e.g., reviewing the terms of service and any updates to those terms), lawyers' virtual document and data exchange platforms should ensure that

---

[18] ABA Comm. on Ethics & Prof'l Responsibility, Formal Op. 482 (2018).

[19] *See, e.g.*, ABA Comm. on Ethics & Prof'l Responsibility, Formal Op. 483 (2018) ("Even lawyers who, (i) under Model Rule 1.6(c), make 'reasonable efforts to prevent the . . . unauthorized disclosure of, or unauthorized access to, information relating to the representation of a client,' (ii) under Model Rule 1.1, stay abreast of changes in technology, and (iii) under Model Rules 5.1 and 5.3, properly supervise other lawyers and third-party electronic-information storage vendors, may suffer a data breach. When they do, they have a duty to notify clients of the data breach under Model Rule 1.4 in sufficient detail to keep clients 'reasonably informed' and with an explanation 'to the extent necessary to permit the client to make informed decisions regarding the representation.'").

[20] *See, e.g.*, ABA Comm. on Ethics & Prof'l Responsibility, Formal Op. 01-422 (2001).

[21] Pennsylvania recently highlighted the following best practices for videoconferencing security:

- Do not make meetings public;
- Require a meeting password or use other features that control the admittance of guests;
- Do not share a link to a teleconference on an unrestricted publicly available social media post;
- Provide the meeting link directly to specific people;
- Manage screensharing options. For example, many of these services allow the host to change screensharing to "Host Only;"
- Ensure users are using the updated version of remote access/meeting applications.

Pennsylvania Bar Ass'n Comm. on Legal Ethics & Prof'l Responsibility, Formal Op. 2020-300 (2020) (citing an FBI press release warning of teleconference and online classroom hacking).

documents and data are being appropriately archived for later retrieval and that the service or platform is and remains secure. For example, if the lawyer is transmitting information over email, the lawyer should consider whether the information is and needs to be encrypted (both in transit and in storage).[22]

5. Smart Speakers, Virtual Assistants, and Other Listening-Enabled Devices

Unless the technology is assisting the lawyer's law practice, the lawyer should disable the listening capability of devices or services such as smart speakers, virtual assistants, and other listening-enabled devices while communicating about client matters. Otherwise, the lawyer is exposing the client's and other sensitive information to unnecessary and unauthorized third parties and increasing the risk of hacking.

6. Supervision

The virtually practicing managerial lawyer must adopt and tailor policies and practices to ensure that all members of the firm and any internal or external assistants operate in accordance with the lawyer's ethical obligations of supervision.[23] Comment [2] to Model Rule 5.1 notes that "[s]uch policies and procedures include those designed to detect and resolve conflicts of interest, identify dates by which actions must be taken in pending matters, account for client funds and property and ensure that inexperienced lawyers are properly supervised."

a. Subordinates/Assistants

The lawyer must ensure that law firm tasks are being completed in a timely, competent, and secure manner.[24] This duty requires regular interaction and communication with, for example,

---

[22] *See, e.g.*, ABA Comm. on Ethics & Prof'l Responsibility, Formal Op. 477R (2017) (noting that "it is not always reasonable to rely on the use of unencrypted email").

[23] As ABA Formal Op. 477R noted:

> In the context of electronic communications, lawyers must establish policies and procedures, and periodically train employees, subordinates and others assisting in the delivery of legal services, in the use of reasonably secure methods of electronic communications with clients. Lawyers also must instruct and supervise on reasonable measures for access to and storage of those communications. Once processes are established, supervising lawyers must follow up to ensure these policies are being implemented and partners and lawyers with comparable managerial authority must periodically reassess and update these policies. This is no different than the other obligations for supervision of office practices and procedures to protect client information.

[24] The New York County Lawyers Association Ethics Committee recently described some aspects to include in the firm's practices and policies:

- Monitoring appropriate use of firm networks for work purposes.
- Tightening off-site work procedures to ensure that the increase in worksites does not similarly increase the entry points for a data breach.
- Monitoring adherence to firm cybersecurity procedures (e.g., not processing or transmitting work across insecure networks, and appropriate storage of client data and work product).
- Ensuring that working at home has not significantly increased the likelihood of an inadvertent disclosure through misdirection of a transmission, possibly because the lawyer or nonlawyer was distracted by a child, spouse, parent or someone working on repair or maintenance of the home.

associates, legal assistants, and paralegals. Routine communication and other interaction are also advisable to discern the health and wellness of the lawyer's team members.[25]

One particularly important subject to supervise is the firm's bring-your-own-device (BYOD) policy. If lawyers or nonlawyer assistants will be using their own devices to access, transmit, or store client-related information, the policy must ensure that security is tight (e.g., strong passwords to the device and to any routers, access through VPN, updates installed, training on phishing attempts), that any lost or stolen device may be remotely wiped, that client-related information cannot be accessed by, for example, staff members' family or others, and that client-related information will be adequately and safely archived and available for later retrieval.[26]

Similarly, all client-related information, such as files or documents, must not be visible to others by, for example, implementing a "clean desk" (and "clean screen") policy to secure documents and data when not in use. As noted above in the discussion of videoconferencing, client-related information also should not be visible or audible to others when the lawyer or nonlawyer is on a videoconference or call. In sum, all law firm employees and lawyers who have access to client information must receive appropriate oversight and training on the ethical obligations to maintain the confidentiality of such information, including when working virtually.

b. Vendors and Other Assistance

Lawyers will understandably want and may need to rely on information technology professionals, outside support staff (e.g., administrative assistants, paralegals, investigators), and vendors. The lawyer must ensure that all of these individuals or services comply with the lawyer's obligation of confidentiality and other ethical duties. When appropriate, lawyers should consider use of a confidentiality agreement,[27] and should ensure that all client-related information is secure, indexed, and readily retrievable.

7. Possible Limitations of Virtual Practice

Virtual practice and technology have limits. For example, lawyers practicing virtually must make sure that trust accounting rules, which vary significantly across states, are followed.[28] The

---

- Ensuring that sufficiently frequent "live" remote sessions occur between supervising attorneys and supervised attorneys to achieve effective supervision as described in [New York Rule of Professional Conduct] 5.1(c).

N.Y. County Lawyers Ass'n Comm. on Prof'l Ethics, Formal Op. 754-2020 (2020).

[25] *See* ABA MODEL REGULATORY OBJECTIVES FOR THE PROVISION OF LEGAL SERVICES para. I (2016).

[26] For example, a lawyer has an obligation to return the client's file when the client requests or when the representation ends. *See, e.g.*, MODEL RULES OF PROF'L CONDUCT R. 1.16(d). This important obligation cannot be fully discharged if important documents and data are located in staff members' personal computers or houses and are not indexed or readily retrievable by the lawyer.

[27] *See, e.g.*, Mo. Bar Informal Advisory Op. 20070008 & 20050068.

[28] *See* MODEL RULES OF PROF'L CONDUCT R. 1.15; *See, e.g.*, ABA Comm. on Ethics & Prof'l Responsibility, Formal Op. 482 (2018) ("Lawyers also must take reasonable steps in the event of a disaster to ensure access to funds the lawyer is holding in trust. A lawyer's obligations with respect to these funds will vary depending on the circumstances. Even before a disaster, all lawyers should consider (i) providing for another trusted signatory on trust

lawyer must still be able, to the extent the circumstances require, to write and deposit checks, make electronic transfers, and maintain full trust-accounting records while practicing virtually. Likewise, even in otherwise virtual practices, lawyers still need to make and maintain a plan to process the paper mail, to docket correspondence and communications, and to direct or redirect clients, prospective clients, or other important individuals who might attempt to contact the lawyer at the lawyer's current or previous brick-and-mortar office. If a lawyer will not be available at a physical office address, there should be signage (and/or online instructions) that the lawyer is available by appointment only and/or that the posted address is for mail deliveries only. Finally, although e-filing systems have lessened this concern, litigators must still be able to file and receive pleadings and other court documents.

III. Conclusion

The ABA Model Rules of Professional Conduct permit lawyers to conduct practice virtually, but those doing so must fully consider and comply with their applicable ethical responsibilities, including technological competence, diligence, communication, confidentiality, and supervision.

accounts in the event of the lawyer's unexpected death, incapacity, or prolonged unavailability and (ii) depending on the circumstances and jurisdiction, designating a successor lawyer to wind up the lawyer's practice.").

Appendix B

Riverside Trial Setting Order

SUPERIOR COURT OF THE STATE OF CALIFORNIA, COUNTY OF RIVERSIDE

CASE TITLE: CASE NO.: DATE: ______________, 2020	Department 1	
PROCEEDING: Amendment to Trial Setting Order		

The Trial Setting Order previously or concurrently issued in this case is hereby amended and supplemented by the provisions set forth below.

This Amendment to Trial Setting Order supersedes any prior Amendment.

____ The trial date previously set is continued or advanced to ________________, 2021, at 8:30 A.M. in Department 1.

1. **VIRTUAL TRIALS ARE A NECESSITY.**
   a. Because of the public health crisis resulting from the COVID-19 pandemic, the Court is not able to welcome counsel, parties, witnesses, or jurors into our civil trial courtrooms for the purpose of in-person civil trials. Nor can the Court predict with reasonable certainty when it will be able to do so. In particular, the Court cannot guarantee that it will be doing so by the time of the date set for the trial in this case.
   b. The Court cannot simply continue trials until the COVID-19 pandemic disappears and the pre-pandemic "normal" is restored. The conditions and practices that were normal at the beginning of 2020 may never return. Even if they were to return 6, 12, or 18 months from now, the limited resources of the Riverside Superior Court would be unable to try the backlog of trials that would accumulate over that period. In short, the Court cannot afford to postpone trials when it has the time and resources to hear those trials now.

c. Accordingly, if the Court is not yet conducting in-person trials by the time that the trial date arrives, that fact by itself will not justify a continuance of the trial. Instead, the trial will proceed as scheduled.
    i. If the trial is to be presented to a jury, the parties shall be prepared to present the jury trial remotely. (See Cal. Rules of Court, emergency rule 3(a)(1) and rule 3.670(f)(3).) If the party or parties who requested a jury trial are not prepared to present the case to a jury remotely, then the Court may find that those parties have forfeited their respective rights to a jury trial, and the trial may proceed as soon as possible thereafter as a bench trial.
    ii. If the trial is to be presented to the Court without a jury, the parties shall be prepared to present the trial to the Court without coming into the courtroom.
        1. If all parties agree, the trial may be conducted by presenting all evidence through stipulated facts, declarations, deposition transcripts, or some combination thereof, and by presenting all opening statements, objections, and arguments remotely.
        2. If all parties are unable to agree to dispense with live testimony, then the trial shall be conducted by presenting the entire trial—all testimony, evidence, opening statements, objections, and arguments—remotely. (See Cal. Rules of Court, emergency rule 3(a)(1) and rule 3.670(f)(3).) If any party is not prepared to do so, then the Court may deem that failure to constitute a failure to appear at trial. If that party is a plaintiff, the Court may dismiss the case. (Code Civ. Proc., § 581, subds. (b)(5) & (*l*).) If the party is a defendant, the Court may bar that defendant from participating in the trial.

d. If the parties can agree upon a referee who will agree to conduct an in-person trial in facilities outside any RSC courthouse, the parties shall file a stipulation for the appointment of such a referee. (Code Civ. Proc., § 638.) Any such stipulation shall be supported by a declaration of the proposed referee, consenting to such an appointment.

2. **AMENDMENTS TO THE TRIAL SETTING ORDER REGARDING VIRTUAL TRIALS**

In the event of a virtual trial, the Trial Setting Order in this case is amended as follows:

a. Those portions of sections A and C of the Trial Setting Order that require personal appearances on the day of trial shall not apply.

b. Notwithstanding section C.3 of the TSO:

   i. All documents required by RSC Local Rule 3401 to be filed on the date of trial, including all exhibits, shall instead be delivered directly to the clerk of this department no later than 48 hours before the day and time on which the trial is scheduled to begin. If the clerk of this department is not available, then those documents shall be delivered to the clerk's office. If the clerk's office is not open, then those documents shall be deposited in the drop box at the courthouse to which the matter is assigned.

   ii. All documents required to be provided to opposing counsel on the day of trial shall be delivered directly to opposing counsel no later than 48 hours before the day and time that the trial is scheduled to begin.

3. **CONDUCTING VIRTUAL TRIALS**

If any portion of the trial is to be presented remotely, all parties shall comply with the provisions set forth below to the extent applicable.

a. **Maintaining the Decorum of the Court**

   Participants appearing remotely must observe rules and procedures related to court appearances, including, without limitation, rules related to attire and the consumption of food or drink during the proceedings. Participants shall eliminate all visual and auditory distractions if appearing remotely.

   Participants appearing in person must adhere to Court, Health Department and/or other government issued requirements in effect at the time addressing the COVID-19 crisis, including, without limitation, wearing of masks and social distancing.

b. **Pre-Trial Arrangements**

   i. **Remote Platform**. For the remote presentation of the entire trial either to a jury or to the Court, the parties shall use Cisco Webex Meetings video. If the only portion of a trial to be presented remotely are the arguments of counsel—for instance, if the trial is on the basis of stipulated facts or written testimony—then the argument may be presented telephonically via Webex, just as is done for the telephonic hearings on the Court's daily law-and-motion and case-management calendar.

ii. **Court Management of Platform**. The Court will manage and control the proceedings, including being designated the "Host" of the video conference, and will exercise control over the various technological settings.

iii. **Preparing to Meet Technological Requirements**. Parties and their counsel appearing remotely shall use best efforts to ensure that there will be clear video and audio transmission during the trial, including adequate familiarity with the designated platform and related software and hardware, *e.g.*, microphones, webcams, headphones, and multiple monitors.

iv. **Adequate Equipment**.

1. The Parties are responsible for ensuring that each attorney, party, witness, and the court reporter appearing remotely is familiar with the designated platform and has the equipment to participate in the remote proceeding without undue delays, including the following: speakers, microphone, webcam, laptop, or monitor. Counsel may wish to ensure that each witness is able to participate in the videoconference with one device (or screen) and review exhibits on another.
2. The Court shall ensure that each juror appearing remotely is familiar with the designated platform and has the following equipment to participate in the remote proceeding without undue delays: speakers, microphone, webcam, laptop, or monitor.

v. **Test Sessions**. In advance of the remote proceeding, at least one test session must be conducted by counsel with each of their witnesses appearing remotely in which the witness practices using the designated platform, becomes familiar with the process for viewing electronic exhibits, and tests all audio and video equipment (including settings) that will be used at trial.

vi. **Court Reporter**. The court reporter may appear remotely. After conference with counsel party and the reporter, the Court will determine whether the reporter will be unmuted for the duration of the proceeding to allow for timely and effective requests for clarification.

vii. **Interpreter**. A court interpreter may appear remotely. After conference with counsel and the interpreter, the court will determine the protocol for those proceedings involving an interpreter appearing remotely for any "Limited English Proficient" (LEP) participant, which may include the following requirements: If appearing remotely for a LEP party, the interpreter will work with counsel to establish a phone bridge connection to facilitate simultaneous interpretation. The LEP party using the interpreter's service must have access to two devices, one for the remote video conferencing software, and a separate phone line. Once the interpreter and LEP are connected via phone bridge, the interpreter will mute themselves from the video conferenced proceeding and interpret simultaneously over the phone. They will unmute themselves in order to interpret short statements or responses from the LEP. When interpreting for a witness, the court will determine if the interpreter will use the consecutive mode, and will not use the phone bridge.

viii. **Camera/Remote Venue Set Up**. The faces of each participant in the trial must clearly be visible while speaking. No masks shall be worn for any participant in the trial appearing remotely. To the extent possible, each participant's webcam should be positioned at face level relatively close to the participant. The background must be neutral.

ix. **Multiple Participants in the Same Room**. Multiple participants in the same room is not allowed. First, to do so the parties would have to be masked. Second, inherent ethical issues of having two people in the same room will undoubtedly arise.

x. **Party Identification**. All participants shall use their full first and last name when signing on to the designated platform.

xi. **Scheduling**. If participants are located in materially different time zones, the proceeding will be scheduled to be reasonably convenient for all, which may require a shorter than normal trial day.

xii. **Confidentiality**. The Parties shall meet and confer in advance of the start of trial regarding a protocol for the use of confidential information, including sealed exhibits, at trial. The Parties shall provide a joint recommendation to the Court before trial commences.

xiii. **Procedure for Sidebars**. The Parties shall meet and confer in advance of the start of trial regarding a protocol for conducting virtual sidebars at trial. The Parties shall provide a joint recommendation to the Court before trial commences.

c. **Public Access and Prohibition of Recording**

i. All trial proceedings will be live-streamed and available for listening by the public. However, the general public must not be provided with access codes provided to court participants.

ii. Unless authorized by the Court, recording of a court proceeding held by video or teleconference, including "screen-shots" or other visual or audio copying of a hearing, is prohibited. However, exhibits may be copied as addressed later. Violations of these prohibitions may result in sanctions, including restricted entry to future hearings, denial of entry to future hearings, removal of Court-issued media credentials, or any other sanctions deemed appropriate by the Court.

d. **Opening Statements and Closing Arguments**

i. Counsel may use the "share screen" function in the designated platform to display demonstratives during openings and closings. Counsel must meet and confer to exchange any visuals or exhibits to be used in the opening statements or closing argument. Any disputes regarding the demonstratives sought to be used by a party shall be addressed with the Court. Counsel may not screen share any visual with the jury unless counsel has stipulated or until the court has ruled on any objection.

e. **Witnesses, Exhibits and Presentation of Testimony**.

i. **Witness Lists**. In addition to the witness list to be filed, the parties shall provide the clerk with a witness list annotated with the email address of each witness.

ii. **Joining the Trial**. Witnesses must access the trial proceedings using the credentials provided by the Court at least 10 minutes before the scheduled start time for their examination. The witness will be directed to the virtual waiting room where he or she will remain until the Court is ready to admit the witness to the virtual courtroom. Nonparty witnesses are only permitted in the virtual courtroom while they are testifying; unless the court orders otherwise, a nonparty witness is not permitted to view or listen to the testimony of other witnesses prior to their testimony.

iii. **Conduct During Testimony**. The attorney calling the witness is responsible for ensuring the witness has a separate video and audio feed. Attorneys should not attempt to "share" a connection with a witness. Witnesses may not have any notes or documents with them at the time of the remote appearance, other than the trial exhibits exchanged by the parties or notes or documents that have been shown to opposing counsel at least twenty-four (24) hours in advance of the witnesses' testimony. Witnesses may not do any research, review any materials, or communicate with anyone else in any manner, including by text, cell-phone, chat, or other means when virtually "on the stand." The Court may require witnesses to back up from their webcam so the Court and counsel can see their hands for the duration or portions of their testimony

iv. **Witness Oath/Affirmation**. In addition to the standard admonitions, before each witness testifies, the Court will ask the witness to affirm: (i) no one else is present in the remote room where the witness is testifying other than those, if any, authorized by the Court; (ii) that all communications with the witness during his or her examination will be on the record, other than communications with the witness and his or her attorney of record during breaks, and (iii) that the witness will not engage in any direct or indirect communications with anyone during his or her examination other than those communications made on the record

v. **Exhibits**.

1. The Parties shall meet and confer in advance of the trial to discuss a protocol for use of exhibits at trial

and present a joint recommendation to the Court at the pretrial conference. The joint recommendation must address the form of exhibits (e.g., electronic and/or paper), acceptable file formats (e.g., .pdf; .doc; .jpeg; .mpeg; *etc.*), how exhibits will be exchanged among the parties, court and presented to jurors (e.g., via email, electronic joint repository, or in-meeting file transfer, *etc.*), and any procedures that may be required for "oversized" exhibits, deposition transcripts, or "non-standard" exhibits of any kind. The Court will give significant weight to the Parties' joint recommendation.

2. If the Parties fail to reach agreement on a joint protocol, the Court will implement a protocol along the following lines, subject to modifications, as appropriate for each matter:
   a. All exhibits to be used on direct and cross examination, except for impeachment, shall be submitted electronically to the Court, all counsel, and the court reporter at a time to be designated by the court, unless otherwise ordered by the trial judge. Each exhibit shall be accessible as an individual document, named electronically according to its exhibit number (*e.g.*, Ex. 1). It is the responsibility of the attorney offering the witness to ensure that the witness has the link to the proceedings and to electronic copies of all exhibits that will be used with that witness, including those of the opposing parties and will be accessible to all jurors whether appearing in person in court or remotely.
   b. **Form**. All exhibits to be used on direct and cross examination, except for impeachment, shall be submitted in tabbed binders to the Court, all counsel, and the court reporter, and/or submitted electronically via jump drive or any other method agreed by the parties. Printed exhibits shall be printed single sided and in black and white, provided that the exhibit shall be printed in color where

reasonably necessary to ascertain its meaning in the context of the proceedings. It is the responsibility of the attorney offering the witness to ensure that the witness has copies in paper form of all exhibits to be used with the witness, including those of the opposing parties, and that those exhibits are available to the witness in the same form that has been provided to counsel and the Court.

c. **Original Documents**. If a true "original" document needs to be entered into evidence, the original should be submitted to the Court as part of its copy of the evidence binder. Such a document should be clearly identified as a true "original."

d. **Impeachment Exhibits**. If counsel wishes to use a document for impeachment purposes that was not previously disclosed as an exhibit, counsel must email an electronic copy of the document to the Court, trial counsel, and the witness at the time counsel seeks to use the document with the witness. In lieu of email, counsel may use the chat function in the designated platform or other platform as designated by the court to send the document to the Court, counsel, and the witness or post the document on a secure document repository that counsel has made available to the Court, counsel, and the witness.

e. **Sealed Paper Copies**. If counsel prefers, he or she may also send a paper copy of the documents that counsel anticipates using for impeachment purposes to the Court and counsel for the other parties at least one business day before the anticipated use of those documents. Two copies of each document must be provided to counsel for the witness with whom the documents will be used. Counsel may enclose the documents in an envelope or box that is sealed and marked DO NOT OPEN UNTIL FURTHER NOTICE. Counsel may package each document in its own sealed envelope so long as it identifies the inner contents in such a way that the recipient can be directed to open a specific envelope. Each envelope shall have a marking across the seal. The envelope shall not be opened unless and until counsel and the witness are instructed to do so by the questioning attorney or the Court. The questioning attorney has the right to request that all recipients return each and every package that they were not authorized to open.

vi. **Objections**. The witness must stop speaking when either counsel objects. After the objection is made, the Court will be the first to speak and will instruct counsel how the Court wishes to proceed.

vii. **Juror Questions**. Before each witness is excused, the Court will request jurors submit any questions they may have for that witness to the judge using the chat or other function. The Court will use a virtual break-out room to discuss the questions posed by the jurors with counsel and shall determine whether the question or a modification of the question will be asked.

f. **Technological Considerations during the Hearing**.

i. **How to Join**. Each attorney, witness, party, and juror who plans to attend any portion of the trial will be a "case participant" and will receive login credentials from the Court. Case participant login information is not public and must not be shared with anyone other than counsel of record and the Court. Everyone who is not a "case participant" will be an "attendee" (*i.e.*, a member of the press or public). Attendees will be able to hear the court proceeding by live stream from the court's website and will not be able to participate in it beyond observing. As stated above, nonparty witnesses may not see or hear the court proceeding until they are called to testify.

ii. **Chat Features**. The chat function allows participants to type text (comments) during the proceeding and also allows participants to send files to other participants. The Court will determine whether the use of the chat function will be allowed during trial after discussion with the Parties. If allowed, the Court will enable the chat function for case participants only and the following rules shall apply: (i) counsel may not initiate *ex parte* "chats" with the Court; (ii) counsel may not "chat" with a witness at any time while the witness is "on the stand" for any purpose unless authorized by the Court. Except for juror questions submitted to the Court via the "chat" feature, messages relayed through the "chat" feature will not become a part of the Court record unless ordered by the Court. Documents transmitted through the chat feature, such as a document to be used for impeachment purposes, will be made part of the Court record but the text of the message transmitting them will not. If counsel transmits a document through the chat feature, counsel must so state on the record and must identify the document for the record and ensure that the court reporter has a copy of it.

iii. **Break-out Rooms**. The Court may use virtual break-out rooms for "side bar" discussions with counsel during trial, to discuss proposed juror's questions to witnesses, for witness waiting rooms, for hearings outside the presence of the jury, or for any other proceedings as needed. The Court will determine if discussions that take place in the break-out rooms are to be made part of the record.

iv. **Addressing Technological Difficulties**. If a participant is disconnected from the videoconference or experiences some other technical failure, the participant shall use best efforts to promptly re-establish the connection and shall take no action which threatens the integrity of the proceeding (*e.g.*, communications with a third party related to anything other than resolving the technical issue). If the connection cannot be re-established within approximately five minutes, the Court may take steps to "pause" the trial, which may include moving participants into the virtual waiting room or one or more separate break-out rooms, at which time counsel shall meet and confer in good faith to develop a joint proposal regarding how to proceed. If the Court deems it unfair to any Party to continue the proceedings because of a technical failure, the Court may postpone or terminate the videoconference at any time and take such other steps as may be necessary to ensure the fairness and integrity of the proceedings.

g. **Jury Deliberations and Discharge**

i. After determining each sitting juror's ability to appear remotely, and consideration of all other relevant factors, the Court will determine whether jurors will be required to appear in person, appear remotely or be allowed to appear either in person or remotely during jury deliberations.

ii. Upon discharge of the jury, the Court shall provide instructions to remotely-appearing jurors regarding notes jurors may have taken during the course of the trial. These instructions may include an instruction that juror notes taken during the trial are to be destroyed.

4. **AMENDMENTS TO THE TRIAL SETTING ORDER REGARDING MANDATORY SETTLEMENT CONFERENCES**

a. Until further notice, any mandatory settlement conference will be conducted remotely, either by telephone or by video. Therefore, those portions of sections A and G of the Trial Setting Order that require personal appearance of counsel, the parties, and any insurance professional at the MSC are suspended. Instead, all such individuals shall participate in the MSC remotely. (See Cal. Rules of Court, emergency rule 3(a)(1) and rule 3.670(f)(3).)

b. The MSC will generally be set in Department 12.

c. The MSC may be set in Department 1, the department to which it assigned for trial.

i. If both sides agree to allow Judge Riemer, the presumptive trial judge, to conduct the settlement conference, the parties shall file a Stipulation to Allow Trial Judge to Conduct Settlement Conference that has been executed by both the parties and their counsel no later than seven days prior to the MSC. (RSC local form RI-CI028, available on the Court's website at www.riverside.courts.ca.gov/FormsFiling/LocalForms/ri-ci028.pdf.)

ii. If the parties do not agree to the settlement conference being conducted by Judge Riemer, the parties shall notify the Court of that decision to allow the Court the opportunity to arrange for a different judicial officer to conduct the MSC. That notice shall be filed as soon as possible prior to the MSC, and in no event later than seven days in advance.

Craig G. Riemer, Judge of the Superior Court
v. 8

Appendix C

Recommended Questions for Online Jury Selection or Online Jury Service

Conditions for Online Jury Selection

You may be asked to participate in jury selection remotely from home or work using a videoconferencing platform.
Once the jury is selected, jurors will serve in person at the Courthouse [with appropriate public health measures in place, including required wearing of masks and socially distanced seating in the courtroom.] Please answer the following questions to help the Court determine if your conditions are appropriate for participation in remote jury selection.

Q1 Do you have access to a quiet and private space and without interruption or distractions between the hours of 9 a.m. to 5 p.m. Monday through Friday, when Court will be in session?

- ◯ Yes
- ◯ Not Sure
- ◯ No

Q2 Do you have any medical or physical limitiations that would make you unable to participate in a group conversation with the the Court and attorneys for 1.5 to 2 hours?

- ◯ Yes
- ◯ Not Sure
- ◯ No

Q2a Please explain your potential limitations:

Q3 What type of internet service do you have at that quiet and private location? (Check all that apply.)

- ☐ Dialup
- ☐ Broadband (high speed) such as cable, fiber optic or DSL
- ☐ Cellular data plan or hot spot
- ☐ Satellite
- ☐ Other
- ☐ Don't Know

Please explain the kind of Internet service you have:

[ ]

Q4 Which of the following types of electronic devices with video cameras do you own or have substantial access to? (Please check all that apply.)

- ☐ Desktop with webcam or embedded camera
- ☐ Laptop with webcam or embedded camera
- ☐ Tablet or iPad
- ☐ Smartphone (for example, an iPhone, Samsung Galaxy, or Samsung Note)
- ☐ None of the above

Q5 What kind of telephone service do you have access to? (You may need to use a separate telephone line to connect to the trial.) Please check all that apply.

- ☐ Cell Phone service
- ☐ Traditional landline
- ☐ Internet phone
- ☐ None

Q6 Do you rely primarily on cell service (rather than wi-fi) for Internet access?

- ○ Yes
- ○ No

Q7 Are you limited in the number of minutes or the data usage that you could use for jury service?

- ○ Yes
- ○ Not Sure
- ○ No

Q8 Who is your cell service provider?

- ◯ Verizon
- ◯ AT&T
- ◯ T-Mobile
- ◯ Sprint
- ◯ US Cellular
- ◯ Other

Please enter the name of your cell service provider:

[ ]

Q9 Remote jury selection is carried out in groups of 15 jurors in approximately two-hour blocks. We require you to be available between the hours of 9 am and 5 pm on the day of your service, but closer to the time jury selection starts, we will inform you of the time block in which you are scheduled. Generally, your questioning would be completed within that time block but you may be asked to return for additional individual questioning. Would you be physically able to participate in the court proceedings using a desktop or laptop with webcam or embedded camera; a tablet or iPad; or a Smartphone during such a schedule?

- ◯ Yes
- ◯ No

Q10 Is there any reason that you could not participate in online jury selection,engaging in conversation with the Court and attorneys, as needed. in a private and quiet place and without interruption or distractions?

- ◯ Yes
- ◯ No

Please explain your answer:

[ ]

Conditions for Online Jury Service

You may be asked to serve as a juror remotely from home or work, or as an online juror using a videoconferencing platform. Please complete the following questions to help the Court determine if your conditions are appropriate for online jury service.

Q11 Do you have access to a quiet and private space and without interruption or distractions from 9 a.m. to 5 p.m. Monday through Friday, when Court will be in session?

- ○ Yes
- ○ Not Sure
- ○ No

Q12 Do you have any medical or physical limitiations that make you unable to watch or listen to an online proceeding for about 1 to 1.5 hours a time without a break?

- ○ Yes
- ○ Not Sure
- ○ No

Q12a Please explain your potential limitations:

[ ]

Q13 What type of internet service do you have at that quiet and private location? (Check all that apply.)

- ☐ Dialup
- ☐ Broadband (high speed) such as cable, fiber optic or DSL
- ☐ Cellular data plan or hot spot
- ☐ Satellite
- ☐ Other
- ☐ Don't Know

Please explain the kind of Internet service you have:

[ ]

Q14 Which of the following types of electronic devices with video cameras do you own or have substantial access to? (Please check all that apply.)

- ☐ Desktop with webcam or embedded camera
- ☐ Laptop with webcam or embedded camera
- ☐ Tablet or iPad
- ☐ Smartphone (for example, an iPhone, Samsung Galaxy, or Samsung Note)
- ☐ None of the above

Q15 What kind of telephone service do you have access to? (You may need to use a separate telephone line to connect to the trial.) Please check all that apply.

- ☐ Cell Phone service
- ☐ Traditional landline
- ☐ Internet phone
- ☐ None

Q16 Do you rely primarily on cell service (rather than wi-fi) for Internet access?

- ○ Yes
- ○ No

Q16a Are you limited in the number of minutes or the data usage that you could use for jury service?

- ○ Yes
- ○ Not Sure
- ○ No

Q16b Who is your cell service provider?

- ○ Verizon
- ○ AT&T
- ○ T-Mobile
- ○ Sprint
- ○ US Cellular
- ○ Other

Please enter the name of your cell service provider:

[ ]

Q17 Trials generally run from 9 a.m. until 5 p.m. on weekdays, with several breaks, including a lunch break, during the proceedings. Would you be physically able to watch and listen to court proceedings using a desktop or laptop with webcam or embedded camera; a tablet or iPad; or a Smartphone during such a schedule?

- ◯ Yes
- ◯ No

Q18 Is there any reason that you could not serve as an online juror, view and listen to the trial from 9:00 a.m. until 5:00 p.m. in a private and quiet place and without interruption or distractions?

- ◯ Yes
- ◯ No

Please explain your answer:

Appendix D

Online Proceeding Questionnaire

1. Which of the following best describes your participation in the online court proceeding?
   - ❑ Judge
   - ❑ Courtroom Staff
   - ❑ Attorney
   - ❑ Juror
   - ❑ Witness
   - ❑ Litigant or Party
   - ❑ Other
2. In what type of proceeding did you participate?
   - ❑ Hearing
   - ❑ Bench Trial
   - ❑ Jury Trial
3. In what state or territory was the trial you participated in? (List States)
4. Did you have a computer, tablet, or device sufficient to participate in an online hearing or trial?
   - ❑ Yes
   - ❑ No
5. Did you have internet access to participate in a remote/online hearing or trial?
   - ❑ Yes
   - ❑ No
6. In general, how strong or weak was your internet connection during the proceeding or trial?
   - ❑ Very strong
   - ❑ Somewhat strong
   - ❑ Somewhat weak
   - ❑ Very weak

7. How frequently or infrequently did you encounter disruptions in connectivity resulting in poor visual images or sound?
   - ❑ Very frequent
   - ❑ Somewhat frequent
   - ❑ Somewhat infrequent
   - ❑ Very infrequent
8. How easy or difficult was it for you to see the participants (judge, attorneys, witnesses, jurors) during the hearing or trial?
   - ❑ Very easy
   - ❑ Somewhat easy
   - ❑ Somewhat difficult
   - ❑ Very difficult
9. How easy or difficult was it for you to clearly see the documents, exhibits, or demonstrative aids during the hearing or trial?
   - ❑ Very easy
   - ❑ Somewhat easy
   - ❑ Somewhat difficult
   - ❑ Very difficult
10. How easy or difficult was it for you to manage the computer equipment you were using for the hearing or trial?
    - ❑ Very easy
    - ❑ Somewhat easy
    - ❑ Somewhat difficult
    - ❑ Very difficult
11. How easy or difficult was it for you to use the videoconferencing platform during the hearing or trial?
    - ❑ Very easy
    - ❑ Somewhat easy
    - ❑ Somewhat difficult
    - ❑ Very difficult
12. How easy or difficult was it for you to stay attentive during the hearing or trial?
    - ❑ Very easy
    - ❑ Somewhat easy
    - ❑ Somewhat difficult
    - ❑ Very difficult
13. How comfortable or uncomfortable was your physical environment during the hearing or trial?
    - ❑ Very comfortable
    - ❑ Somewhat comfortable
    - ❑ Somewhat uncomfortable
    - ❑ Very uncomfortable

14. How engaged did you feel you were during the hearing or proceeding?
    - ❑ Very engaged
    - ❑ Somewhat engaged
    - ❑ Somewhat disengaged
    - ❑ Very disengaged
15. How focused or distracted were you during the hearing or trial?
    - ❑ Very focused
    - ❑ Somewhat focused
    - ❑ Somewhat distracted
    - ❑ Very distracted
16. If you were an attorney, how easy or difficult was it for you to present evidence and arguments in an online hearing or trial?
    - ❑ Very easy
    - ❑ Somewhat easy
    - ❑ Somewhat difficult
    - ❑ Very difficult
    - ❑ I am not an attorney

    If difficult, to what do you attribute the difficulty?
17. If you were a juror, how easy or difficult was it for you to deliberate with other jurors?
    - ❑ Very easy
    - ❑ Somewhat easy
    - ❑ Somewhat difficult
    - ❑ Very difficult
    - ❑ I was not a juror
18. If you were an attorney or juror, how easy or difficult was it for you to communicate with the court during the hearing or trial?
    - ❑ Very easy
    - ❑ Somewhat easy
    - ❑ Somewhat difficult
    - ❑ Very difficult
    - ❑ I was neither an attorney nor juror
19. Have you ever served as a juror for an in-person trial conducted at a courthouse?
    - ❑ Yes
    - ❑ No
    - ❑ I was not a juror

    If yes, what did you feel were the benefits of participating in an online trial compared to your previous in-person experience?

    If yes, what do you feel were the problems with participating in an online trial compared to your previous in-person experience?

20. How convenient or inconvenient was it for you to participate in an online/remote hearing or trial?
    - ❑ Very convenient
    - ❑ Somewhat convenient
    - ❑ Somewhat inconvenient
    - ❑ Very inconvenient
21. If you were a juror, given the choice in the future, which location would you prefer to serve as a juror?
    - ❑ Courtroom
    - ❑ Online
    - ❑ No preference
    - ❑ I was not a juror
22. If you were called to serve as a juror in the future, how likely or unlikely would you be to report for jury duty in a courtroom?
    - ❑ Very likely
    - ❑ Somewhat likely
    - ❑ Somewhat unlikely
    - ❑ Very unlikely
23. If you were called to serve as a juror in the future, how likely or unlikely would you be to report for jury duty in an online forum using a platform like Zoom?
    - ❑ Very likely
    - ❑ Somewhat likely
    - ❑ Somewhat unlikely
    - ❑ Very unlikely
24. Please rate your satisfaction or dissatisfaction with your experience participating in an online trial.
    - ❑ Very Satisfied
    - ❑ Somewhat satisfied
    - ❑ Neither satisfied nor dissatisfied
    - ❑ Somewhat dissatisfied
    - ❑ Very dissatisfied
25. What recommendations do you have for improving online/remote hearings and trials?

Demographics

We are asking for this information to assess the representation of different groups in the litigation and trial process.

26. What is your gender?
    - Male
    - Female
    - Non-binary
    - Other: ______________
    - Decline to say
27. What is your age?
    - 18–25
    - 26–35
    - 36–45
    - 46–55
    - 56–65
    - 66–75
    - Over 75
    - Decline to say
28. What is your ethnicity?
    - African American/Black
    - Latino/Latina/Hispanic
    - Asian/Pacific Islander
    - Native American
    - Middle Eastern/Arab
    - White/Caucasian
    - Mixed Ethnicity/Other: ______________
    - Decline to say
29. What is your average yearly household income?
    - Under $25,000
    - $25,000–$50,000
    - $51,000–$75,000
    - $76,000–$100,000
    - $101,000–$150,000
    - Over $150,000
    - Decline to say

30. What is your highest level of completed education?
    - High school or less
    - Some college
    - AA degree
    - Bachelor's degree
    - Master's degree
    - Post-graduate degree
    - Professional or trade school degree
31. Do you normally care for young children, disabled or differently abled, or seniors during the day?
    - Yes
    - No
32. What best characterizes the work or activities you do during the day?
    - Employed full time
    - Employed part time
    - Self-employed
    - Unemployed
    - Homemaker
    - Disabled/differently abled
    - Retired
    - Student
    - Decline to say
33. What is your job title? ________________________________
    - Decline to say

Appendix E

Ideal Features in a Web Platform Customized for Courtroom Use

By the Online Courtroom Project

While many platforms are able to meet the basic functions of the hearing room or courtroom, none of the platforms we have tested, nor any of the other web-conferencing systems we have reviewed or considered, are ideal for courtroom use in their present forms. In his closing comments during the OCP's June, 2020 online trial demonstration, Richard Gabriel referred to it as an "architecture issue," and we believed that is an apt metaphor. "Rooms" on these platforms, like rooms in your house, are designed for a particular function. But there is no other communication function in existence that is quite like a trial. Zoom is designed for meetings or for webinars. Other platforms are designed for focus groups or for research mock trials. Actual trials have some commonalities with all of these settings, but some key differences as well.

Based on our experiences, we have reached the conclusion that none of the platforms will be fully satisfying without significant customization. What seems most realistic would be for the companies in this space to develop either a bespoke version of their service for courtroom use, or to build-in substantially greater flexibility to create that experience by switching features on and off.

While a detailed punch-list would necessarily be specific to the design and capabilities of each platform, in general terms, a courtroom-customized online conferencing system should include the following features:

- **Role Differentiation**: Instead of just host, presenter, and attendee, we need specialized logins and interfaces based on trial role. A person would sign in as judge, counsel, support team, party, witness, court personnel, juror/venire, or gallery viewer. These roles should be identified and grouped on-screen. Each of those roles would have different levels of access and use programmed into the system.

- **Chat Channels**: Parties and the court should have a message or "chat" feature that is visible only to the judge, counsel, and court personnel.
- **Lobby**: There should be a designated room or "lobby" where witnesses wait until they are called.
- **Side-Bars**: The judge, counsel, and (if used) court reporter should be able to immediately create a side-bar conference in a single step, without moving or being visible to other participants.
- **Juror's Room**: Jurors should have a designated waiting room where they can get to know each other and chat about anything they want (other than the case, presuming pre-deliberation is not allowed) when court is not in session or when the trial is being conducted out of the jury's hearing.
- **Team Room**: Each party should have an easily accessible room where attorneys, clients, and their trial support teams can meet with full confidentiality when trial is not in session.
- **Phased Screen-Share**: Document sharing should include a system for selective screen-sharing for exhibit review and authentication. A document, photograph, or video should be able to be viewed by the court, counsel, and a witness before being admitted and published to the jury.
- **Anonymous Messages**: Jurors should have a system by which one can anonymously send a question to the judge, so that the judge can decide whether that question can be asked of a witness. That system could also be used for personal requests (such as a break), or (without anonymity) to report technical issues.
- **Deliberation Polling**: During deliberations, jurors should have the ability (if they choose to use it) to take anonymous votes on verdict questions, as they would with paper slips if they were in person.
- **Deliberation Exhibit Manipulation**: Jurors should have a system allowing them to easily use and display exhibits to each other during deliberations, and view and mark up a verdict form. When screen-sharing, they should still have the ability to see every other juror.
- **A "Technical Bailiff"**: One individual should have partial access from the jury's deliberation room, such that the bailiff can be summoned by the jurors if needed, and be alerted when the jury has reached a verdict, or when there is another issue (e.g., a juror

loses a connection, or the jury is taking a break or finishing for the day). This bailiff should not otherwise be able to see or hear the content of the jury's deliberations.

- **Gravitas**: The user interface should convey some of the symbols that mark the courtroom as a special and solemn place: It shouldn't look like just another web-conference. As much as possible, the visual and functional interface should appeal to the symbols that we associate with the socially vital functions of a courtroom.

Appendix F

Online Resources

The following documents are available electronically via the Online Courtroom Project website:

1. Online Courtroom Project—Publications and Manuals. URL: https://www.onlinecourtroom.org/publications-and-manuals Includes links to the following:

    Example Civil Trial Online Supplemental Juror Questionnaire from King County Superior Court, State of Washington.

    Demonstration Questionnaire: Online Version of the Written Juror Questionnaire for *State of Minnesota v. Derek Chauvin*.

    Several other reports and manuals relating to online practice during the pandemic.

2. Updated list of articles relevant to online courtrooms. URL: https://www.onlinecourtroom.org/ocpresources
3. Online Jury Trials: OCP's Summary and Recommendations Report. URL: https://www.onlinecourtroom.org/demonstration-report